一场肉眼看不见的战争

A Battle between Bacteria and Antibiotics

沈建忠 张嵘 主编

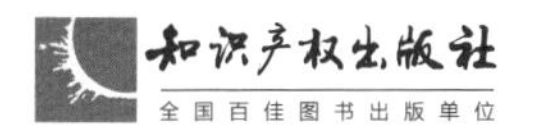

图书在版编目（CIP）数据

细菌与抗生素之战：一场肉眼看不见的战争（汉英对照）/ 沈建忠，张嵘主编. —北京：知识产权出版社，2017.7

ISBN 978-7-5130-4841-5

Ⅰ. ①细… Ⅱ. ①沈… ②张… Ⅲ. ①细菌—青少年读物 ②抗菌素—青少年读物 Ⅳ. ① Q939.1-49 ② R978.1-49

中国版本图书馆 CIP 数据核字（2017）第 065867 号

内容提要

本书以汉英双语注释搭配彩色插画的形式，旨在通过图文并茂的形式普及和推广细菌与抗生素的知识。本书对中小学开展抗菌药物合理使用与细菌耐药科普教育与宣传活动，以及帮助青少年从小树立抗菌药物合理使用观念具有重要意义。

责任编辑：刘晓庆　　　　**责任出版**：刘译文

细菌与抗生素之战：一场肉眼看不见的战争（汉英对照）

XIJUN YU KANGSHENGSU ZHIZHAN：YICHANG ROUYAN KANBUJIAN DE ZHANZHENG

沈建忠　张　嵘　主编

出版发行：知识产权出版社有限责任公司	网　　址：http://www.ipph.cn http://www.laichushu.com
电　　话：010-82004826	
社　　址：北京市海淀区气象路 50 号	邮　　编：100081
责编电话：010-82000860 转 8073	责编邮箱：396961849@qq.com
发行电话：010-82000860 转 8101	发行传真：010-82000893
印　　刷：北京科信印刷有限公司	经　　销：各大网上书店、新华书店及相关专业书店
开　　本：710mm × 1000mm　1/16	印　　张：11.25
版　　次：2017 年 7 月第 1 版	印　　次：2017 年 7 月第 1 次印刷
字　　数：180 千字	定　　价：38.00 元

ISBN 978-7-5130-4841-5

本书由国家重点基础研究发展计划（973计划）项目（编号：2013CB127200）“畜禽重要病原菌抗生素耐药性形成、传播与控制基础性研究”的资助出版。

参编人员名单

顾丹霞　李佳萍　胡付品　汪　洋　凌卓人　杨　璐

欧嫣然　马藤菲　史晓敏　沈应博　胡燕燕　池　丹

林晓伟　林　迪　孙巧玲　王汉宇　黄子衔　李金樾

周维平　王　斌　俞鸿达

校译人员名单

Joy Tan　沈张奇　朱　奎　王少林

前　言

抗生素作为20世纪人类最伟大的发现之一，在感染性疾病的防控和治疗中发挥着举足轻重的作用，然而随着它在临床和农业中的广泛应用，抗生素耐药问题日益严峻，不但严重威胁公众的健康，也给全球的经济可持续发展造成了沉重的负担。抗生素耐药问题的背后，一个非常重要的原因就是普通人群对抗生素的合理使用存在很大误区。我们国家作为人口大国，既是抗生素生产大国，也是抗生素使用大国，因此我们在推动抗生素合理使用有效对抗抗生素耐药的道路上责无旁贷。在2016年杭州G20峰会前夕，国家卫生和计划生育委员会、国家发展和改革委员会等14个部门联合印发了《遏制细菌耐药国家行动计划（2016—2020年）》，从国家层面提出了细菌耐药防控工作的主要措施，着重提到的重要措施是将抗菌药物合理使用的相关知识纳入中小学健康教育内容并尽快加以落实。

抗菌药物的合理使用要从孩子抓起，为了让更多细菌与抗菌药物的知识从高校的象牙塔走向中小学生，中国农业大学和浙江大学医学院师生以及中学生联手奉上了《细菌与抗生素之战——一场肉眼看不见的战争》。这本科普读物用中英两种语言、趣味生动的文字和唯美的画面浅显易懂地阐述了自然界的微生物、抗生素的发展，以及细菌耐药性形成的原因，让不同年龄段的孩子都能从中获取知识，使得合理使用抗生素的观念深入人心，从而推动合理使用、拒绝滥用抗生素的全民行动的不断深入。

遏制细菌耐药性的发展，我们在行动！

中国工程院院士

中国农业大学教授

倪建强

2017年6月18日

Preface

Antibiotics, as one of the most significant discoveries in the 20th century, play an essential role in the prevention and treatment of infectious diseases. However, with the widespread application of antimicrobials in clinical treatments and agricultural use, the problem of antimicrobial resistance (AMR) has become increasingly severe. It not only poses a great threat to public health, but also has become a heavy burden to the sustainable development of the global economy. One of the prime causes accounting for AMR is the misunderstanding of antibiotic treatments among the general public. China, with its enormous population, produces as well as consumes a large amount of antibiotics. Accordingly, we need to shoulder the responsibility of promoting the appropriate use of antibiotics to prevent AMR from triggering more undesirable consequences.

On the eve of the G20 Hangzhou Summit, the National Health and Family Planning Commission of the People's Republic of China, along with the National Development and Reform Commission and other 14 departments, issued the *National Action Plan of China for The Containment of Antimicrobial Resistance (2016—2020)* which proposed a series of countermeasures, especially spreading relevant knowledge concerning the rational use of antibiotics into elementary and secondary education. This knowledge should no longer be confined to university students, but permeate into children's daily lives. In order to achieve this goal, a popular-science book entitled "*Invisible War: A Battle between Bacteria and Antibiotics*" was published jointly by China Agricultural University's and Zhejiang University, School of Medicine. Through lively pictures and vivid narratives both in Chinese and English, this book provides a brief introduction about microorganisms, antibiotics and the emergence of AMR for children of various age groups. It is expected to enhance the awareness of rational antibiotic use among both children and adults, to avoid abuse of antibiotics in society.

To curb the spread of AMR, we are in action!

Academician of Chinese Academy of Engineering

Professor at China Agricultural University

Jianzhong Shen

18th June, 2017

目　录

Content

第一章

无“微”不至的细菌

Bacteria Exist Everywhere

1. 微生物大家族

你知道微生物是什么吗？微生物是大自然中一大群外形很小、结构简单的微小生物。微生物通常在生活中直接用眼睛是看不到的，必须在显微镜下放大几百倍甚至几十万倍才能观察到（但某些真菌，如蘑菇是可以直接用眼睛看到的）。自然界中有很多不同的微生物，它们分布在各个地方，与人类的关系密切。

如果将一个完整的细胞比作一个人，那么细胞壁和细胞膜就像人类的皮肤，细胞质就像人类的血液，

1. The Microorganisms Family

Do you know what microorganisms are? They are a bunch of tiny creatures in nature with extremely simplified structures. Apart from some fungi such as mushrooms that are large enough to be seen, most microorganisms are invisible to the naked eye. To observe them, we must use a microscope to magnify their original sizes by hundreds or even hundreds of thousands of times. There are many different microorganisms in nature.Their presences are distributed out in various places and they are closely related to human beings.

If we were to compare a cell to a human, a cell's wall and membrane are like a person's skin.Cytoplasm is like

细胞器就像人类的各个器官，而细胞核则充满了遗传物质，使子细胞遗传母细胞的特征。

那么，你知道微生物是如何分类的吗？微生物按细胞大小、细胞结构、细胞组成等可以分成三大类，即原核细胞型、真核细胞型和非细胞型微生物。

1.1 原核细胞型微生物

让我们先来看看原核细胞型微生物吧！原核细胞型微生物没有完整的细胞核，只有类似细胞核的核质，而且它们具有不完善的器官，

blood and cellular organelles are just like various kinds of organs in the human body. The cell's nucleus is full of genetic materials, which can be passed down from the parent cells to offsprings.

How are all these microorganisms organized and classified? They are classified according to their sizes, structures and cellular components. Microorganisms are classified into three major groups, prokaryotic microorganism,eukaryotic microorganism and non-cellular microorganism.

1.1 Prokaryotic Microorganism

First, let's take a look at prokaryotic microorganisms. They only have a nucleoid instead of intact nucleus.

核糖体是体内唯一的器官。原核细胞型微生物包括细菌、支原体、衣原体、立克次体、螺旋体和放线菌等。

其中，细菌数量最大、种类最多，它们体型微小，结构简单。机智的科学家们在显微镜下观察发现，有的细菌是球形，像一个足球，我们称之为球菌，如肺炎链球菌、金黄色葡萄球菌；有的是杆状，称之为杆菌，如大肠杆菌、白喉棒状杆菌；有的是螺形，称之为螺形菌，如幽门螺杆菌。

细菌都有细胞壁、细胞膜、细胞质和细胞核，而有些细菌还有自

What's more, their cell organelles areincomplete and the only organelle existing are ribosomes. Prokaryotes include bacteria, mycoplasma, chlamydia, rickettsia, helix, actinomycetes and etc.

Bacteria have the most amount of variation in species, and they are also the largest in numbers. Bacteria are tiny microorganisms with a simple structure. Outstanding scientists observed them under the microscope and discovered that some are spherical, which look like footballs, and we called them coccus. *Streptococcus pneumonia* and *Staphylococcus aureus* are examples of coccus. Others are rhabditiform named bacillus, such as *Escherichia coli* and *Corynebacterium diphtheria*. Some

己的“法宝”，它们有特殊的结构，包括保护自己不受伤害的荚膜和芽胞，帮助自己向前运动的鞭毛，以及促使自己黏附宿主的菌毛（图 1.1）。

支原体是最小的原核细胞型微

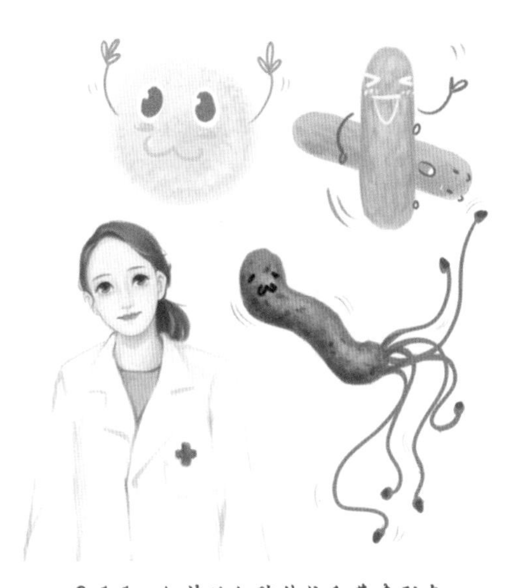

图 1.1　细菌的细胞结构和基本形态
Figure 1.1　The bacterial cell structures and basic forms

are spiral shaped, called spirilla, such as *Helicobacter pylori*.

Bacteria are composed of cytoderm, cytomembrane, cytoplasm and nucleoid. Some bacteria even have their own special structures, including capsule and spore, which can protect themselves from harm. The flagella help push themselves forward to move around, and fimbriae can help bacteria adhere to a host (Figure 1.1).

Mycoplasma is the smallest prokaryote. Without cytoderm, they can transform into various shapes just like plasticine, such as sphericity, rhabditiform, filiform, ramose and etc. An interesting fact, mycoplasma can grow on mediums prepared by scientists, and they look like delicious

生物。支原体没有细胞壁，所以它们可以像橡皮泥一样形态多变，有球形、杆状、丝状和分支状等。有趣的是，支原体在科学家们配制的培养基上长大后看上去就像一个个美味的“荷包蛋”（图1.2）。

衣原体和立克次体比较“娇气”，学不会独立自主，就像寄生虫一样，必须寄生在其他真核细胞内才能存活（图1.2）。衣原体的发育比较特殊。科学家们观察发现，衣原体有两种不同的形态，一种称为原体，一种称为始体。一个原体可以转变成很多始体，始体又可以转变成等量的原体。

fried eggs when they are fully grown (Figure 1.2).

Just like parasites, chlamydia and rickettsia must live by other eucaryota since they are rather weak, therefore they cannot grow independently (Figure1.2). However, chlamydia goes through a special life cycle. Two different patterns are observed by scientists during the development of chlamydia, they are the elementary body (EB) and initial body (RB). EB is infectious and can easily infect the host cells. After entering into the host cells, EB can transform into RB, and then RB develops into EB, infecting new host cells to start another cycle.

Spirochetes and actinomycete both grow at a slow rate. Spirochetes are

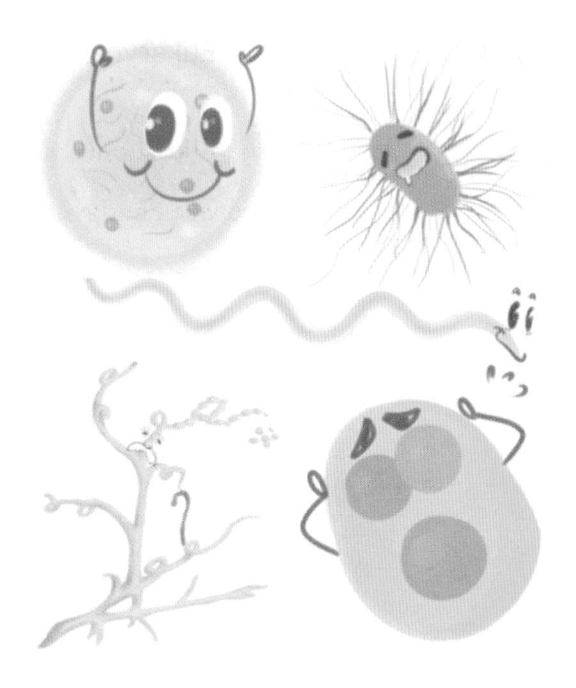

图 1.2　支原体，衣原体，立克次体，螺旋体和放线菌
Figure 1.2　Mycoplasma, chlamydia, rickettsia, helix and actinomycetes

螺旋体和放线菌生长缓慢。其中，螺旋体细细长长，却十分活泼，常常弯曲成螺旋状；而放线菌呈丝状或树枝般的分枝状（图 1.2）。

spindly in shape and extremely lively, often curling into a spiral. On the other hand, actinomycetes present filiform or dendroid like branches (Figure 1.2).

1.2 真核细胞型微生物

真核细胞型微生物的细胞核和细胞器都比较完整。真菌就属于这类微生物。真菌的形态多种多样，小到直接用眼睛看不到的新生隐球菌，大到如我们平常吃的美味的香菇、木耳等。所以按形态、结构，真菌又可以分为单细胞真菌和多细胞真菌。单细胞真菌呈圆形或椭圆形，而多细胞真菌由菌丝和孢子组成（图1.3）。真菌不“挑食”，在不同营养成分的培养基上都能生长，但有些真菌生长缓慢。

1.2 Eukaryotic Microorganism

Eukaryotic microorganisms' organelles and neucleus are relatively complete and intact. Take fungi as an example, they come in a variety of sizes.The smallest fungus is *Crypto-coccusneoformans* which can't be seen by the naked eye, while some large fungi are edible, such as mushroom and agaric.Fungi can be classified to unicellular fungi and multicellular fungi according to their structures.The first one is either round or oval in shape, the latter are composed of hyphae and spores (Figure 1.3). Fungi are not "picky" with their food, they are able to survive in all kinds of environment, but some grow very slow.

1.3　非细胞型微生物

非细胞型微生物是结构最简单并且个体最小的微生物。它们没有完整的细胞结构，也不能独立存活，

1.3 Noncellular Microorganism

Noncellular microorganism is a group of microorganisms that have the simplest structure and smallest in size. They do not have a complete celluar structure, neither can they live

图 1.3　真菌的细胞结构，多细胞真菌的基本结构
Figure 1.3　The fungal cell structure, the basic architecture of multicellular fungi

只能在其他活细胞内生长，就好像寄居蟹一样，赖着不走。病毒就是这类微生物。

一个成熟的病毒颗粒称为病毒体。各种病毒体的大小差别悬殊。大部分病毒呈球形或近似球形，少数病毒呈杆状、丝状、砖块状等。病毒没有完整的细胞结构。它们的基本结构是由核心和衣壳构成的核衣壳。核心类似细胞核，充满病毒的遗传物质，衣壳保护核心不受伤害。有些病毒的核衣壳外还有包膜（图 1.4）。

independently.Instead, they can only survive within other viable cells, just like a pagurian. The virus is a type of noncellular microorganism.

A mature virus particle is known as the virus body. Viruses have a variety of different sizes, and most viruses are spherical shaped or near-spherical, while a few of them are rod-shaped, filiform-shaped, or brick-shaped and etc. Virus does not have an integrated structure, with its nucleocapsid being the basic structure, it is composed by core and capsid. The core is similar to a cell nucleus and it is packed with genetic material. The capsid protects the core against all harm and danger. Some virus also has an extra structure called envelope outside the nucleocapsid. (Figure 1.4).

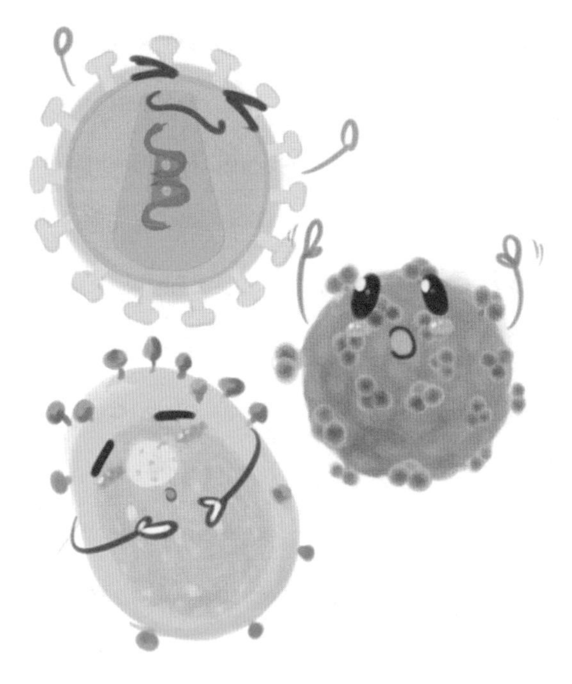

图 1.4 各类病毒的基本形态
Figure 1.4 The basic form and shape of different of viruses

2. 微生物大家族的重要成员之一——细菌

2. An Important Member of Microorganism Family—Bacteria

微生物是如此庞大的大家族，可能地球上所有人加在一起也没有

Microorganisms have such a large family, where its number of

一种细菌的数量惊人。妈妈肚子里的胎儿体内几乎是没有细菌的，但他（她）一旦离开妈妈的肚子，呱呱坠地，各种各样形形色色的微生物就开始寻找适宜的环境在宝宝身上"安居乐业"了。除了暴露在空气中的皮肤，与外界相通的像口腔、上呼吸道、肠道、泌尿系统、生殖系统等部位，都寄居着各种数量惊人的微生物。

以口腔中的细菌为例，一个人的口腔中到底有多少细菌呢？科学家们发现，一个清洁的口腔，每颗牙齿表面有 1 000~100 000 个细菌；而一个不怎么注意清洁的口腔，

members significantly exceeds the total population of human beings on earth. Human infants lives in a sterile environment within the mother's womb before being born, but they interact with and get inhabited by various kinds of bacteria after being born. Apart from their skin, which is directly exposed to the air, a surprising number of bacteria also inhabit in our oral cavity, upper respiratory tract, intestine and the urogenital tract.

Let's take the oral cavity bacteria for example. How many bacteria live in a person's mouth? According to researchers, the number of bacteria existing in a well-cleaned oral cavity ranges from 1000 to 100,000; for someone who does not pay attention

每颗牙齿表面的细菌数量可以高达10亿个。

有趣的是，同样是寄居在口腔中的细菌却有着不同的种族和群体。它们在口腔中有各自的生存空间。有的喜欢群居在两颊内，有的喜欢群居在舌腹，有的喜欢群居在舌背，而厌氧菌更喜欢群居在牙齿缝隙之间。这些不计其数的小家伙构成了丰富多彩的细菌王国。

细菌是微生物大家族中最重要的成员，可以根据特殊的染色方法，按照细菌"外衣"细胞壁的厚薄把它们分为革兰氏阳性菌（紫色）和革兰氏阴性菌（红色）两大类。科

to cleaning his teeth, the number of bacteria on their teeth can reach up to a billion.

Interestingly, even though they all inhabit within the oral cavity, bacteria in your mouth vary greatly in terms of their type and species. They have their own living spaces insides your mouth. Some prefer to inhabit between the cheeks, some like to gather on the ventral surface of the tongue, others enjoy living on the back of the tongue. Anaerobic bacteria particularly like living between tooth gaps. These tiny creatures make up the large and magnificent kingdom of bacteria.

Bacteria are the most important members in the microorganism family,and they can be classified

学家们把细菌放在显微镜下观察，按照形状的差异又大体把细菌分为球菌、杆菌和螺旋菌。细菌大家族也是有“族谱”的哦！科学家们根据亲缘远近对它们进行划分，从远到近依次分为界、门、纲、目、科、属、种。当然，属于同一“种”的细菌亲缘关系较近。

细菌大家族中的成员们除了拥有细胞膜、细胞壁、细胞质和核区这些常见的细胞结构外，还拥有一些特殊的法宝。这些结构是它们运动、生长和生存所必须的。下面就把细菌的看家本领向大家一一道来。

in several ways. Bacteria can be divided into two major groups by Gram stain: Gram-positive bacteria (colored purple) with thick cell walls and Gram-negative bacteria (colored red) with thin cellwalls. Additionally, they can also be distinguished by their shapes,such as cocci, bacillus and spirillum, under a microscope. The bacteria family has a family tree! Sicnetists use bacteria's genealogy to classify them according to their relatedness. Starting from: Kingdom, Phylum, Class, Order, Family, Genus and Species. Bacteria belong to the same species share a closer relationship.

Other than common structures such as cell membrane, cell wall, cytoplasm

菌毛

在很多革兰氏阴性杆菌和少数革兰氏阳性球菌的细胞表面存在一些细短而直硬的丝状物，称为菌毛。菌毛直径为3~7nm，长度为0.5~6μm，而有些菌毛可长达20μm。

菌毛类型很多，科学家们根据功能把它们分成两大类——普通菌毛和性菌毛。普通菌毛能够帮助细菌牢牢地黏附在其他细胞（包括人的呼吸道、消化道和泌尿道的上皮细胞）或物体的表面。有的能使细菌吸附在红细胞上，从而引起红细胞凝集；而性菌毛在细菌的遗传物

and the nuclear region, bacteria also have special appendages which enable them to move, grow and survive. We will talk about the special characteristics of bacteria below.

Pili

Many Gram-negative bacilli and a few Gram-positive cocci have short, hard and straight hair-like figures called pili. The diameter of a pilus is about 3~7 nm, and the length is about 0.5~6 μm, but some can grow up to 20 μm.

There are many types of pili, and scientists divided them into two major categories according to their functions—common pilus and sex pilus. Common

质传递过程中起着极其重要的作用。比如一株细菌想把它的耐药基因和毒力基因馈赠给其他细菌的时候，性菌毛就会在两个细菌间搭起一个“运输管道”，然后把有用的东西运送过去（图 1.5）。

pilus can help bacteria adhere to other cells firmly (including the human epithelial cells of the respiratory tract, digestive tract and urinary tract) or adhere to the surface of the object. Some can assist the bacteria to adhere to the erythrocyte, causing agglutination of erythrocyte. Sex pilus plays an extremely important role in the process of transferring genetic material. For example, if one bacterium wants to gift the resistance genes and virulence genes to another, sex pilus will act as a pipeline to transfer useful information between the two bacteria (Figure 1.5).

图 1.5　细菌通过性菌毛运输物质
Figure 1.5　Transfer of genetic materials by sex pilus

鞭毛

某些细菌菌体上的细长而弯曲的丝状物，称为鞭毛（图 1.6）。鞭毛的长度比菌毛长，是细菌的运动器官。具有鞭毛的细菌大多是弧菌、杆菌和个别球菌。根据鞭毛的着生部位的不同，可将鞭毛分为周生鞭毛、侧生鞭毛和端生鞭毛。鞭毛有

Flagella

Some bacteria have long, curving hair-like structures, called the flagella (Figure1.6), flagellum is a sporting organ of the bacteria. Flagella are longer than pili. In most cases, bacteria with flagella are vibrios, bacilli or individual cocci. According to their location, the flagella can be divided into three types: peri-flagella, lateral-flagella, and polar-flagella.

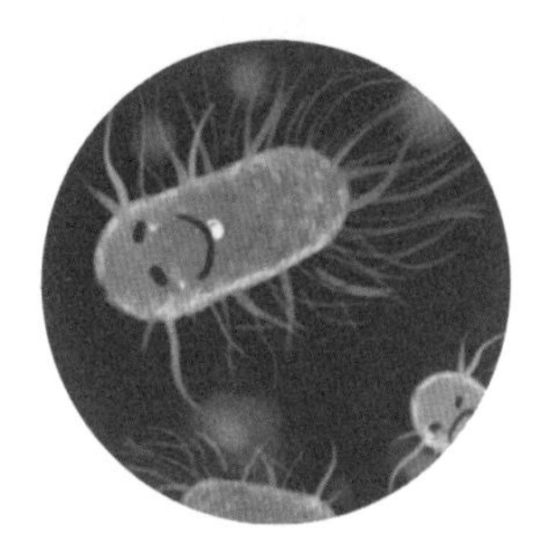

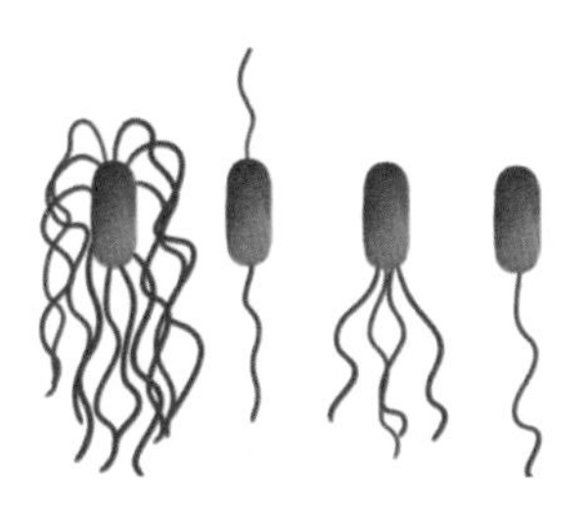

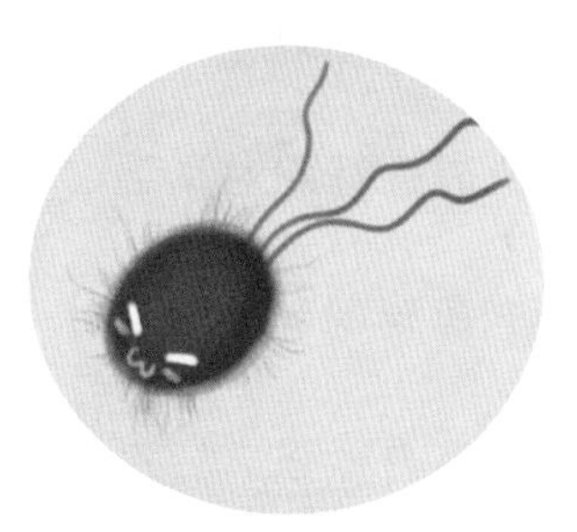

图 1.6 细菌的鞭毛
Figure 1.6 Flagella of bacteria

三种运动方式：在液体中泳动，在固体表面上滑行，以及在液体中旋转梭动。具有鞭毛的细菌可是“运动健将”哦！

Flagella move in through three ways: swimming in liquid, gliding on a solid surface, or rotating and shuttling in liquid. Bacteria with flagella are masters in sports!

芽胞

Spores

一部分细菌非常聪明，当它感觉到环境中没有足够的营养物质维持其生长繁殖的时候，就可以形成芽胞（图 1.7）。芽胞像一个巨大的仓库，使自己进入类似动物的休眠状态。芽胞对高温、紫外线、干燥、电离辐射和很多有毒的化学物质的抵抗力很强。

Some bacteria are extremely clever, when they feel like their environment does not have enough nutrients to support their growth, they can form spores (Figure 1.7). Spores are like giant warehouses, allowing the bacteria to shift into dormacy like other animals. Spores are very strong, they are able to resist against heat, ultraviolet light, dryness, ionizing radiation, and many types of toxic chemicals.

例如，肉毒梭菌的芽胞在沸水中要经过5~9.5小时才被杀死。芽胞的休眠能力更为突出，在常规条件下，一般可以保持几年至几十年不死。据文献记载，有的芽胞甚至可以休眠数百至数千年。最极端的例子是在美国的一块有2500万~

For example, spores of botulinum can only be killed after 5~9.5 hours in boiling water. The spores' dormant ability is outstanding. Generally speaking, they can sleep and live from years to tens of years. According to record, some spore cells can sleep for hundreds to thousands of years. The most extreme case is found in the United States where a piece of amber contained spores that have been

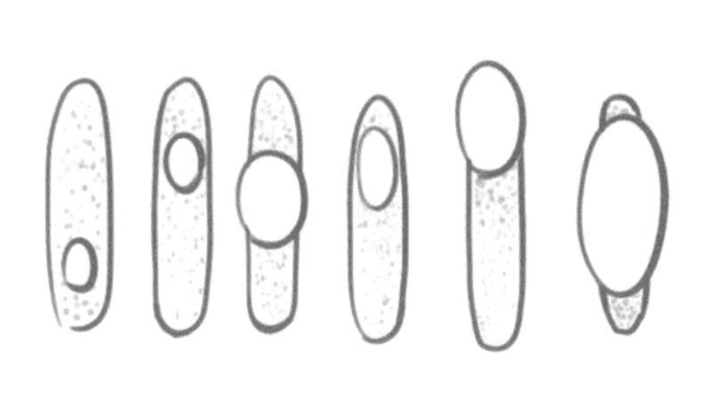

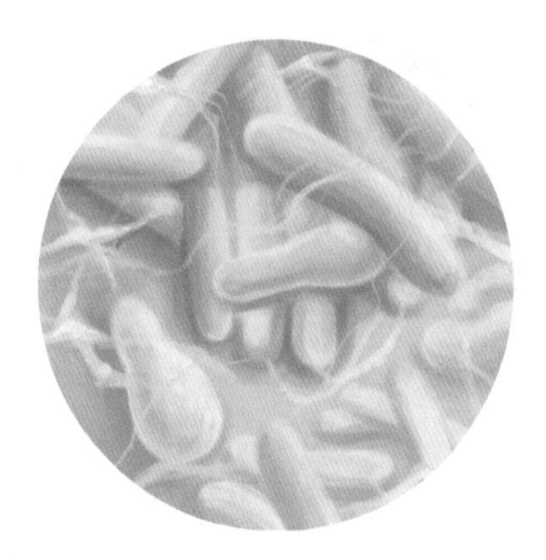

图 1.7　细菌的芽胞
Figure 1.7　Spores of bacteria

4000万年历史的琥珀里，至今从其中蜜蜂肠道内还可以分离到有生命的芽胞。一旦环境中的营养物质满足生长需求，细菌马上由休眠状态复活，这就太可怕了。

alive for 25 to 40 millions of years, you are still able to extract living spores from the bee's intestines. As soon as the nutrients within the environment are sufficient for the spore's growth, it can come back to life immediately. This is amazing!

荚膜

荚膜里某些细菌表面的一层松散的黏液型特殊结构（图1.8）。荚膜的成分主要是葡萄糖与葡萄糖醛酸，也有含多肽与脂质的。荚膜最重要的功能是抗吞噬和黏附作用，使细菌黏附于细胞表面并且躲避"追杀"。革兰氏阴性杆菌和革兰氏阳性球菌均可产生荚膜，如革兰氏

Capsule

Some bacteria have a layer of loose myxoid structure on its surface (figure 1.8), mainly composed of glucose and glucuronic acid, some are composed of polypeptide and lipid as well. The most important functions of capsule are anti-phagocytosis and the adhesion effects, which make the bacteria adhere on the surface of cells to avoid being killed by others. Gram-negative bacillus and Gram-

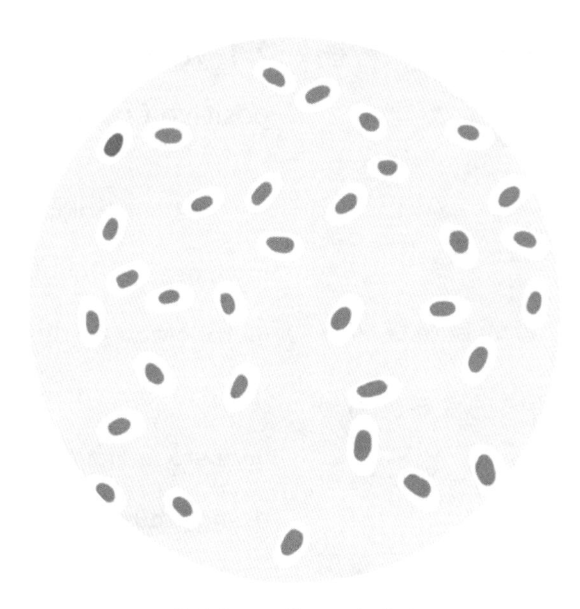

图 1.8 细菌的白色荚膜
Figure 1.8 White capsule of bacteria

阳性球菌的肺炎链球菌和革兰氏阴性杆菌的肺炎克雷伯菌等。

那么，在环境中、人类和动物体内都有哪些代表性的细菌呢？让我们一起来了解一下吧。

positive coccus can produce capsule, such as *Streptococcus pneumoniae* and *Klebsiella pneumoniae* and etc.

What are the typical bacteria living within the environment, human body, and animals? Let's take a closer look.

2.1 革兰氏阳性球菌中的“天使”与“恶魔”

2.1 Angels and Devils of Gram-positive Cocci

葡萄球菌名字的来源非常有趣，就是因为它们在普通显微镜和电镜下像小时候喜欢吃的成串的葡萄（图 1.9）。大多数葡萄球菌不会引起我们和动物生病。例如，表皮葡

The nomination of *Staphylococcus* in Chinese is very interesting, as the morphology of this organism looks like clusters of grapes under the microscope(Figure 1.9). Most of them do not cause disease in human and

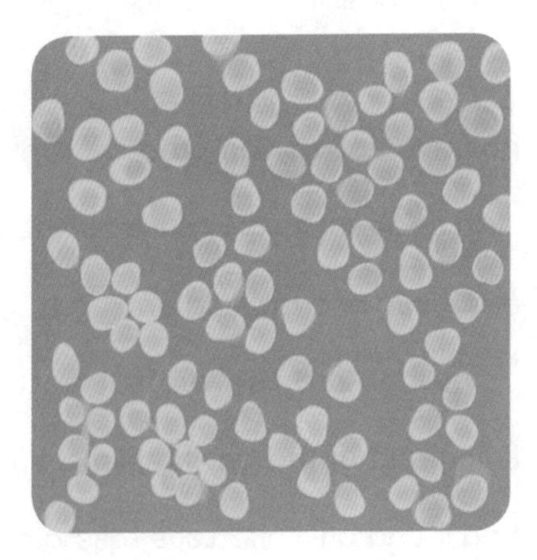

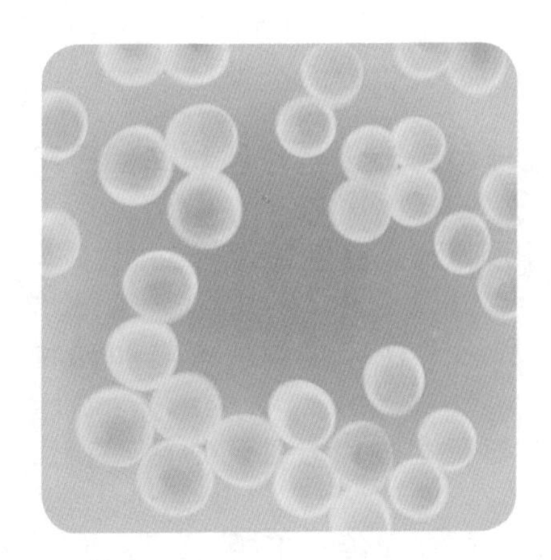

图 1.9 葡萄球菌在显微镜下的形态

Figure 1.9 The morphology of staphylococcus under microscope

萄球菌是皮肤表面的正常细菌，是皮肤的天然屏障。

但当表皮葡萄球菌转移到除皮肤以外的其他地方时，它们可能会使我们生病。例如，它们钻进血液就可引起败血症，钻进心脏就可引起细菌性心内膜炎。金黄色葡萄球菌是葡萄球菌中毒力最强的细菌，会引起人类皮肤化脓性感染、肺炎、败血症、食物中毒、休克等多种可怕的疾病。

链球菌是一群排列成链状的球菌（图1.10）。大多数的链球菌也是正常菌群，广泛分布在自然界、人畜的鼻咽部及粪便中。例如，草

animals.For example, *Staphylococcus epidermidis* is normal flora on the skin and plays a role as natural barriers for our skin.

The growth of *Staphylococcus epidermidis* in the other locations than the skin will cause disease in human. It will cause septicemia and bacterial endocarditis. *Staphylococcus aureus* is the most virulent species among *Staphylococcus*, which could cause a variety of terrible disease, including skin purulent infections, pneumonia, septicemia, food poisoning, shock and et al.

Streptococci are cocci that are grouped together like chains (Figure 1.10). Most of the *Streptococci* are normal groups of floras, and they

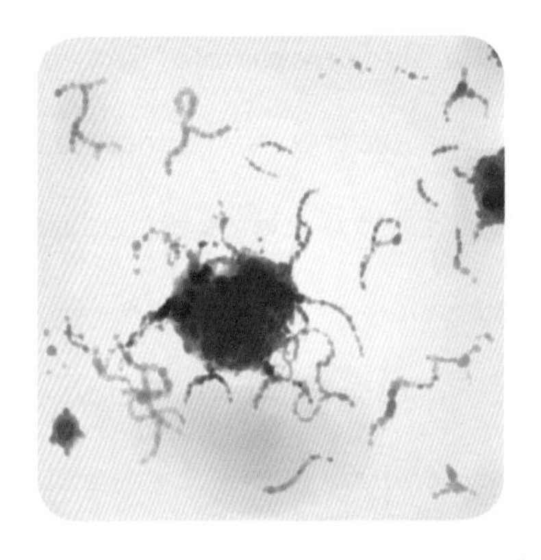
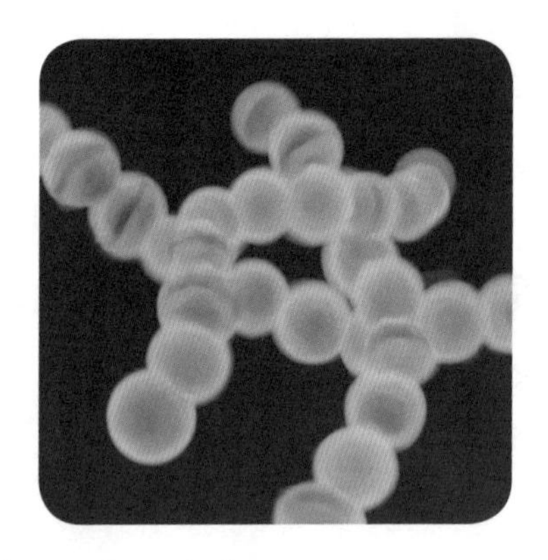

图 1.10 链球菌在显微镜下的形态
Figure 1.10 The morphology of streptococcus under microscope

绿色链球菌常寄居在口腔、呼吸道、消化道等。链球菌中的"恶魔"当属化脓性链球菌和肺炎链球菌，其中化脓性链球菌常引起皮肤、扁桃体等发生化脓性感染和猩红热，而肺炎链球菌常引起脑膜炎、支气管炎和肺炎等疾病，有时候会威胁到我们的生命。

are widely distributed out into the environment,and inside human and animal nasal area and their feces. For example, green streptococcus often inhabits within the oral cavity, respiratory tract, and digestive tract. Streptococcus pyogenes and Staphylococcus pneumoniae are the "devils" of Streptococcus, the former can cause skin and tonsil

肠球菌为圆形或椭圆形、呈单个或成对或短链状排列的革兰氏阳性球菌（图 1.11），无芽胞，无鞭毛。绝大部分肠球菌的生长需要氧

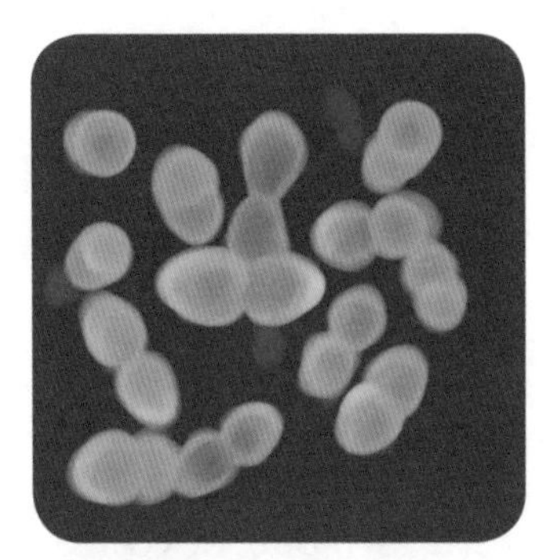
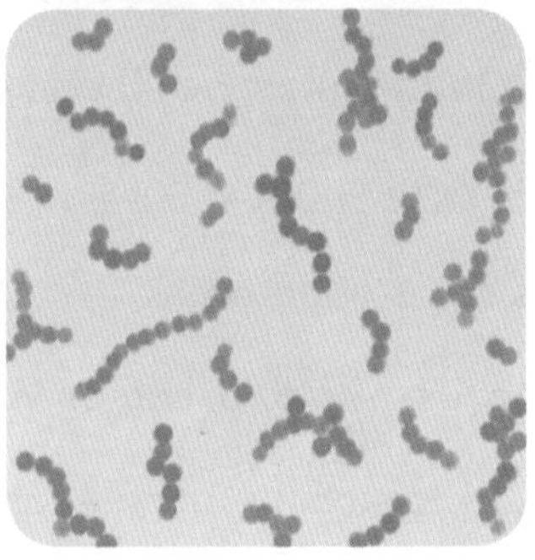

图 1.11 肠球菌在显微镜下的形态
Figure 1.11 The morphology of enterococcus under microscope

pyogenic infections, and it is also the cause of scarlatina; while the latter often causes meningitis, bronchitis, pneumonia and other diseases. These diseases can potentially be fatal.

Enterococci are a kind of round or oval shaped *Gram-positive cocci* that often occur as either single, in pairs, or short chains (figure 1.11). They do not have spores and flagella. Most *Enterococci* need oxygen to survive, and they will die in an environment without oxygen. *Enterococci* are the normal flora of the intestinal tract which can help the body absorb nutrients, but under certain conditions, *Enterococci* may cause infections. *Enterococcus faecalis* and *Enterococcus faecium* being the most common

气，在没有氧气的环境中会死亡。肠球菌是人类和动物肠道的正常菌群，能够帮助肠道吸收营养物质。但是在一定条件下，肠球菌也会造成感染，其中最主要的是粪肠球菌和屎肠球菌，它们可以引起膀胱炎、肾炎和败血症。

2.2 认识革兰氏阴性杆菌

革兰氏阴性杆菌为杆状形态的细菌（图1.12），主要存在于人类和动物的肠道中。由于肠道营养物质丰富，温度适宜，许许多多的细菌喜欢在那里定居。它们像天然的

causes of infection, they can lead to serious cases of cystitis, nephritis and septicemia.

2.2 Gram-negative bacilli

Gram-negative bacilli are shaped like rhabdoids (figure 1.12), and they mainly exist in the intestine of human and animals. Because of the intestine's abundant nutrition and its optimum temperature, many bacteria enjoy settling in the intestine. The intestine is like a natural green house,nurturing various types of bacteria.

The *Escherichia coli* is one of the most commonly found examples. They arecommon florasliving in the intestine, and they play an important

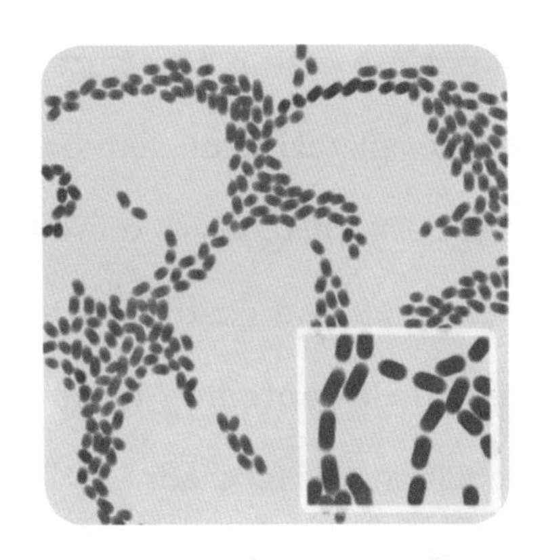
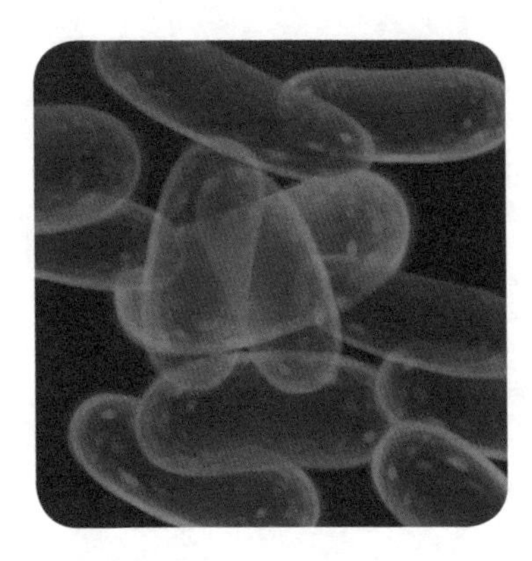

图 1.12 革兰氏阴性杆菌在显微镜下的形态
Figure 1.12 The morphology of Gram-negative bacilli under microscope

温室，孵育了各种细菌。

大肠埃希菌是其中最常见也是最有代表性的成员之一。它是肠道中的正常成员，在维护肠道生态平衡中起着重要的作用。但是当食物和饮用水被污染的时候，大肠埃希菌数量增多，肠道菌群平衡被破坏，某些“破坏分子”

role in maintaining the balance within the intestine. However, when food and water are polluted and the number of *Escherichia coli* increases, the intestinal balance may be broken. In this situation, some subversives invade the intestine and cause gastroenteritis, therefore the count of *Escherichia coli* is used as a pollution index for food and drinking water.

就会趁虚而入引起人类胃肠炎，所以“大肠埃希菌计数”被作为饮水、食品等污染的指标。

还有少数细菌一直以来就会致病，如志贺菌（俗称痢疾杆菌）和沙门菌。它们会引起细菌性痢疾和胃肠炎，严重时甚至会危及生命。

除大肠埃希菌外，肺炎克雷伯菌、奇异变形杆菌、铜绿假单胞菌和鲍曼不动杆菌也是革兰氏阴性杆菌的重要成员。

2.3 讨厌氧气的细菌

有些细菌生长时“厌恶”氧气，

There are also a few bacteria that are completely fatal, such as *Shigella* and *Salmonella*, which cause bacillary dysentery and gastroenteritis. They are life threatening when serious conditions are reached.

Aside from *Escherichia coli*, other bacteria such as *Klebsiella pneumoniae*, *Proteus Mirabilis*, *Pseudomonas Aeruginosa* and *Acinetobacter Baumannii* are also important members of Gram- negative bacilli.

2.3 Anaerobic Bacteria

Some bacteria are named anaerobic bacteria by scientists because of their aversion to oxygen.

科学家们称它们为厌氧性细菌。人体肠道中的厌氧菌，如双歧杆菌，乳酸杆菌等能合成多种人体生长发育必须的维生素，还能利用蛋白质残渣合成必需的氨基酸，并参与糖类和蛋白质的代谢，同时还能促进人体对铁、镁、锌等矿物元素的吸收。这些营养物质对人类的健康有重要作用，一旦缺少会引起多种疾病。

但有些厌氧菌却对人类有害，如引起破伤风的破伤风梭菌就是其中一种厌氧性细菌。它们呈鼓槌状，生命力顽强，在干燥的土壤和尘埃中可存活数年。破伤风梭菌

Anaerobic bacteria living in human's intestine, such as *Bi-fidobacterium* and *Lactobacillus*, can synthesize various kinds of necessary vitamins for our growth and development. They can also take advantage of the protein residues to synthesize essential amino acid, and take part in the metabolism of sugar and protein. Additionally, they can promote the absorption of iron, magnesium, zinc and many other mineral elements. These nutrients are of great importance to our health, people can easily become sick without them.

However, some anaerobic bacteria are harmful to humans, such as *Clostridium tetani*, which are causes to tetanus. They are shaped like drumsticks, and they can live in the dry

个头虽小，却可以夺人生命。当人体受到外伤时，破伤风梭菌最喜欢隐藏在伤口最深处，悄悄地发芽繁殖，释放毒素，使得人体痉挛、抽搐，最后因窒息和呼吸衰竭而死亡。再如艰难梭菌，是引起腹泻的主要病原菌之一。

soil and dust for several years with plenty of vitality. Although *Clostridium tetani* are small size, they can easily take the life of a human. When people get wounded, *Clostridium tetani* are often found hinding in the deepest wounds, silently growing while releasing toxins to the body, which can make people go into convulsion, and eventually suffocate or die from respiratory failure. The *Clostridium difficile* is another example of a troublemaker, which is one of the main pathogens that cause diarrhea.

2.4 人畜共患病原菌

有些细菌偏好寄生在动物身上，由动物传染给我们，引起动物和人类共患病。最可怕的当数鼠疫耶尔森菌，俗称鼠疫杆菌。鼠疫耶尔森菌寄生在野鼠、家鼠身上，会

2.4 Zoonotic Pathogens

Some bacteria prefer the parasitic life on animals, and they make people

引起流行性鼠疫。当大批病鼠死亡后，鼠类身上携带病原菌的鼠蚤转向人类，通过叮咬使人类感染鼠疫耶尔森菌。这些坏蛋可以引起肺鼠疫，病人常常因为咳嗽、胸痛、呼吸困难而死亡。死亡病人的皮肤常常呈黑色，所以又被称为"黑死病"。

此外，它们还可以引起败血症。人类历史上曾发生过三次世界性大流行的鼠疫，每次都引起社会动荡，惨不忍睹。控制老鼠的数量，可以防止鼠疫的发生。新中国成立之后的"除四害"运动成效显著，鼠疫得到了有效控制。

sick through interactions with sick animals, causing zoonotic diseases. The most horrible bacteria are *Yersinia pestis*, which parasitize in rats and mice and cause extreme diseases such as the plague. After the rat dies, rat fleas will carry and spread the dangerous pathogens, Yersinia pestis, to humans by biting them. These horrible bugs can cause pneumonic plague, where the patients often die for coughing, chest pain, and dyspnea. The plague is also known as "black death"because the patient's skin will appear black after death.

In addition, they can also cause sepsis. The world-wide plague pandemic had occurred three times throughout history, and they have always triggered

形形色色的细菌充盈着我们人体和我们所生活的环境，它们对人类、动物、植物的生命活动甚至地球上的物质循环都起着重要的作用。随着我们对细菌王国认识的深入，越来越多的奥秘被揭开，越来越多的细菌为我们所改造而造福人类。

social turbulence. Eliminating rats can prevent the occurrence of the plague. After the establishment of the People's Republic of China, the movement of eliminating the four pests has been effective, and the plague have been effectively controlled.

Bacteria of all shapes and sizes fill our everyday lives and the environment we live in, they play an important role in the life activities of humans, animals and plants, and even for the circulation of materials on earth. As our understanding of the bacterial kingdom broadens, more secrets will unveil, and an increasing number of bacteria are studied and transformed into knowledge that will benefit mankind.

第二章

细菌的发现之旅

The Discovery of Bacteria

1. 细菌王国之门的开启

别看细菌个头小，“来头”可不小。科学家们发现，早在太古宙时期，细菌就已在地球上出现。这些“品味独特”的“小东西”对大气中的氮气有着特殊爱好。它们能够吸取氮气并把它转化成生命体生存所必须的氨基酸，使它们在人类还未诞生就已经在地球上繁衍了很多年。最近，有科学家们惊奇地发现这些细菌曾经一度吃掉了一半大气。这一数据着实令人震惊。

尽管细菌家族历史悠久，但这个神秘的“王国”一直不为人类所

1. The Kingdom of Bacteria

Do not look down upon the bacterium's size, because its origin is a magnificent one. Scientists have found that early in the Archean period, bacteria had already appeared on earth. These "picky-eaters" have special interests in nitrogen within the atmosphere. They can absorb nitrogen and transform it into amino acid, which is a necessity of life, and this enables them to survive for millions of years long before the existence of human beings. Recently, some scientists were surprised to find that these bacteria had eaten half of the atmosphere. This data is both a surprise and a shock.

Although the bacterial family has

知。直到三百多年前，一个叫列文虎克的荷兰人发明了显微镜，才开启了这扇神秘的大门。年轻时，列文虎克是一个对微观世界很感兴趣的人。他没有接受过正规的教育，完全凭借自己的兴趣和毅力磨制镜片，制成了能够放大200倍的显微镜。这是当时最好的显微镜（图2.1）。

借助这一强大的工具，他惊奇地发现在干草浸出液中能观察到许许多多不断蠕动、姿态迥异的“小东西”，它们就是细菌本尊。至此，一个全新的未知世界开始向人类敞开。此后，列文虎克还用显微镜观

a long history, this mysterious kingdom has not been known to mankind for a long time. Until three hundred years ago, a Dutchman named Antonie van Leeuwenhoek invented the microscope, which revealed the mysterious mask of these creatures. The young man was very interested in the microscopic world. Without formal education, his great interest and perseverance lead him to polish his lenses, and he eventually made a microscope capable of magnifying bacteria by two hundred times, which was the best microscope at that time (figure 2.1).

With this powerful tool, he was surprised to find that many "small things" of different shapes were swimming in the leaching waters of

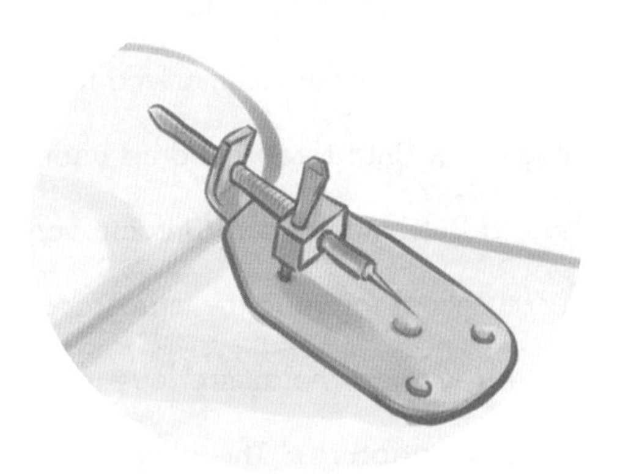

图 2.1　列文虎克自制的显微镜
Figure 2.1　A self-made microscope by Antonie van Leeuwenhoek

察到了口腔中的细菌。他给英国皇家学会写了一封信报告了自己的发现，轰动欧洲。

继列文虎克之后，更多的科学家们前赴后继在细菌王国开辟“新大陆”。不得不提的一个人是法国化学家巴斯德。他虽然不是救死

hay. These are known to be bacteria. The door to an unknown world had been opened by mankind. Later, Leeuwenhoek also observed bacteria from his own mouth with a microscope, and he wrote a letter to the Royal Society and shocked Europe with his discovery.

After Leeuwenhoek, more scientists went on to explore the new world of the bacterial kingdom. Another significant person is a French chemist called Pasteur, although he was not a life-saving doctor, he was the one to confirme the relationship between bacteria and diseases. He also used a series of experiments to refute the prevailing “spontaneous generation” theory (biological

扶伤的医生，却证实了细菌与疾病之间的关系。他还用一系列实验反驳了当时盛行的“自然发生学说”（生物可以从它们所在的物质元素中自然发生，而不是通过上代繁衍产生）。

当然，巴斯德对人类的贡献不仅仅如此，他还创立了“巴氏消毒法”，即用63.5℃的温度加热葡萄酒半小时杀死乳酸杆菌。这个方法既不破坏葡萄酒本身，又解决了葡萄酒变酸的问题，从而挽救了当时岌岌可危的法国酿酒业。

你们知道巴氏消毒牛奶是世界上消耗最多的牛奶品种吗？正因为

creatures can occur naturally from the material elements they are in, rather than through reproduction).

Pasteur's another contribution to mankind is that he created the "pasteurization" process (heating the wine at 63.5°C for half an hour to kill). This solved the problem of *Lactobacillus* wine deterioration without altering the actual taste, and this discovery went on saving the French wine industry.

Did you know that Pasteurized milk is the most consumed milk worldwide? Benefiting from pasteurization, we can now enjoy various kinds of delicious and nutritious milk. But it is worth noting that this method is not a panacea, pasteurized milk usually needs to be

有巴氏消毒法，我们才能享受到这么多美味又营养的牛奶。但值得注意的是，这种方法并不是万能的。巴氏消毒牛奶通常需要在4℃低温储存，这样才能抑制细菌生长，从而防止牛奶变质。

差不多与巴斯德同时代，还有一位著名的德国科学家——科赫。科赫在微生物病原菌方面的卓越贡献，使他获得了1905年的诺贝尔医学和生理学奖。他从患炭疽病的牛身上分离到炭疽杆菌，并把这种菌接种到小鼠身上，使小鼠相互感染患上炭疽病，最后发现从小鼠身上分离的病原菌与最初牛身上分离

stored at 4°C temperature,which can prevent bacterial growth and prevent milk from deterioration.

Among Pasteur's peers, there was a famous German scientist——Koch, his work in microbial pathogen was outstanding, and he went on winning the 1905 Nobel Prize in Physiology and Medicine. He isolated *Anthrax* from a cow infected with anthracnose, and transferred it into mice to infect the mice with anthrax,where he found that the pathogen isolated from mice was identical to the pathogen from the cattle. Through using scientific experimental methods, he was the first to confirm that anthrax is the pathogen of anthrax.

Later on, Koch also found the cause

的病原菌完全一样。就这样，他用科学的实验方法首次证实了炭疽杆菌是炭疽病的病原菌。

后来，科赫还发现了肺结核的罪魁祸首——结核分支杆菌和引起霍乱的霍乱弧菌，并提出了著名的“科赫法则”，为科学家们攻克这些“元凶”引起的传染病打下了基础。

在他的引领下，科学家们后来还发现了白喉棒状杆菌、伤寒杆菌、鼠疫耶尔森菌、痢疾杆菌等可以引起人类大规模感染的病原菌。这一时代成为发现病原菌的“黄金时代”。此外，他还创立了一系列包

of tuberculosis——*Mycobacterium* tuberculosis, as well as *Vibrio cholera* which is the cause of cholera. He proposed the famous "Koch rule", which led many scientists into overcoming causes of infectious diseases.

Under his leadership, scientists later discovered pathogens that cause large-scale infections including *Corynebacterium diphtheriae, Salmonella typhi, Yersinia pestis* and *Shigella dysenteriae*. This era became the "golden age" for the discovery of pathogens. Additionally,Koch created a series of microbiological methods, including the techniques of isolation, and culture, medium technology and dyeing technology. These discoveries brought human's understanding of

括分离纯培养技术、培养基技术和染色技术等微生物学方法，使人们对这些“小东西”的认识进入一个全新的境界。1982年，为了纪念科赫发现结核病原菌100周年，我国还专门发行了一枚纪念邮票（图2.2）。

bacteria onto a whole new level. In 1982, China specially issued a collection of commemorative stamps to celebrate the 100th anniversary of the discovery of *Mycobacterium tuberculosis* (Figure 2.2).

Another significant milestone in the long history fighting against

图 2.2 我国发行的科赫发现结核杆菌一百周年纪念邮票

Figure 2.2 100 years anniversary of the discovery of Mycobacterium tuberculosis (commemorative stamps published by China Post)

纵观细菌研究的历史长河，英国外科医生李斯特提出的消毒术是细菌发现之旅上的又一大里程碑。李斯特在长期临床工作中接触到了很多伤口化脓和感染的病人。他发现闭合性骨折（发生骨折但伤口不暴露）的病人一般不会发生化脓，而开放性骨折（发生骨折且伤口暴露）的病人即使伤势轻微，也会发生化脓。于是，他推断伤口和空气中的细菌是罪魁祸首，并开创了外科消毒术。它大大降低了外科手术病人的感染死亡率。

在数不胜数的科学家们的不断探索下，细菌这些在我们身边但却

bacteria, is disinfection proposal by a British surgeon Joseph Lister. In his clinical practice, Lister observed that many patients suffered from suppurated infections. He noticed that closed fractures (fracture without exposure the wound to air) usually do not lead to suppuration. On the other hand, even the smallest fracture (wound exposed to air) will result in suppuration. Therefore, Doctor Lister deducted that bacteria are the real cause of infection and suppuration. Furthermore, he invented a disinfection method which greatly lowered the incidence of death by infection in surgical operations.

Through the continuing efforts of many scientists, we began to

一直不为我们所知的“邻居”越来越为我们所认识、利用和改造。

2. 区分细菌的方法

科学家们发现细菌后，就开始采用各种各样的方法来研究这些数量惊人的“小东西”。就像人类一样，细菌虽然个头小，但也是要吃东西才能长大。科学家们为了让它们成长，配制了各种各样营养美味的培养基来培养它们。大部分细菌并不“挑食”，在含有羊血的血培养基上长势喜人（图 2.3）。但是一些特别“娇贵”的细菌需要更多的营养。

gradually learn and understand these lovely "neighbors" who live among our lives, where we learned to utilize and modify bacteria to the benefits of humankind.

2. Methods to Distinguish Various Bacteria

After the discovery of bacteria, scientists started to figure out various ways to study these creatures which are tiny but incredible in amount. Although varying in their body sizes, bacteria need foods to grow just like babies do. In order to grow bacteria, scientists designed different kinds of nutritious meals as mediums for

比如，嗜血杆菌需要在特别添加营养素的巧克力培养基（没有添加巧克力哦）上才能生长（图2.4）。

大多数细菌对空气中的氧气和二氧化碳没有特殊要求，而有些细菌喜欢多呼吸点二氧化碳。例如，

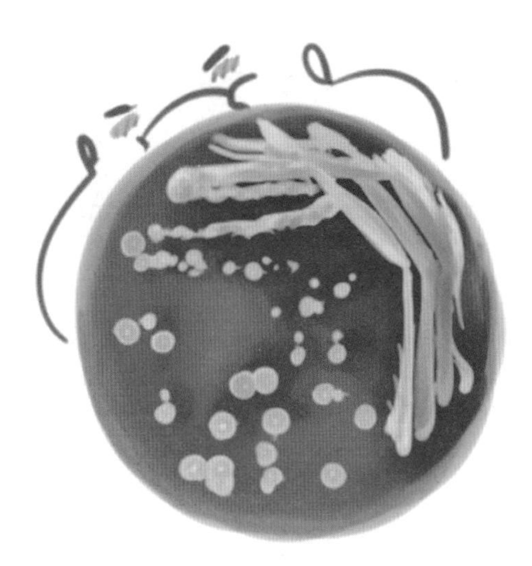

图2.3 细菌生长图——血培养基
Figure 2.3 Bacteria culturing on blood medium

them to grow. Most bacteria are not "picky-eaters". They can grow in a healthy and quick manner on sheep blood(figure 2.3). However, certain kinds of bacteria are rather fragile, and they need more nutrition to survive. For example, *Haemophilus* must be grown on chocolate medium(without containing actual cocoa chocolate) with additional nutrients (figure 2.4).

The majority of bacteria have no special demand for oxygen or carbon dioxide. While some of them, such as *Neisseria Meningitidis* and *Brucella*, prefer a carbon dioxide environment, they feel much more at ease in a high carbon dioxide concentration condition. Another sort of peculiar bacteria extremely dislike carbon

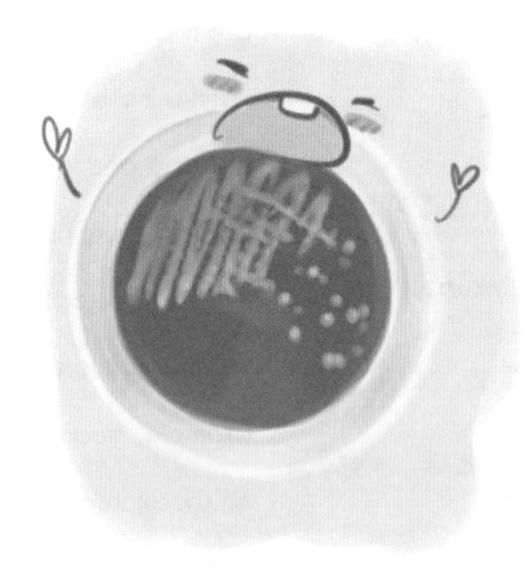

图 2.4　细菌生长图——巧克力培养基
Figure 2.4　Bacteria culturing on chocolate medium

脑膜炎奈瑟氏菌和布鲁氏菌偏爱高浓度的二氧化碳。还有一类“性情怪异”的细菌，极度厌恶有氧环境。对它们而言，氧气就是毒药。例如，隐藏在伤口最深处的破伤风梭菌。它们能在缺氧环境中迅速壮大

dioxide. To them, air is poison. For example, *Clostridium Tetani*, they hide in the deepest part of a wound where they reproduce rapidly and pose serious threats to human's health. But once exposed to air, *Clostridium Tetani* will die soon.

If you think that just because

队伍，危害人类的健康，但一旦暴露在空气中就很快死去。

如果你以为细菌这么小又没有四肢肯定不会动，那就大错特错啦！在载玻片上滴上生理盐水，借助显微镜，我们会观察到它们在“泳池”中欢快游动的身影（图2.5和图2.6）。除了特别大的真菌肉眼可见外，其他细菌个体微小，只能通过显微镜观察，所以显微镜在其中起着特别重要的作用。

但很快科学家们就发现，直接观察会有一个问题，不知道看到的是细菌，还是别的如细胞、气泡或者环境中的杂质。那到底该怎样

bacteria have no legs therefore unable to move, then you are wrong! By placing a drop of saline on a microscope slide, we will be able to observe bacteria "swimming" happily. (Figure 2.5 and Figure 2.6). Apart from bacteria such as fungus which is visible to the naked eye, most bacteria are extremely small in size, they can only be seen through a microscope. Therefore, microscope plays a fundamental role in bacterial research.

Scientists eventually realized that directly observing bacteria had its problems. For example, it was hard to tell cell, bubbles, bacteria and other environmental contaminants apart from each other. So how do you

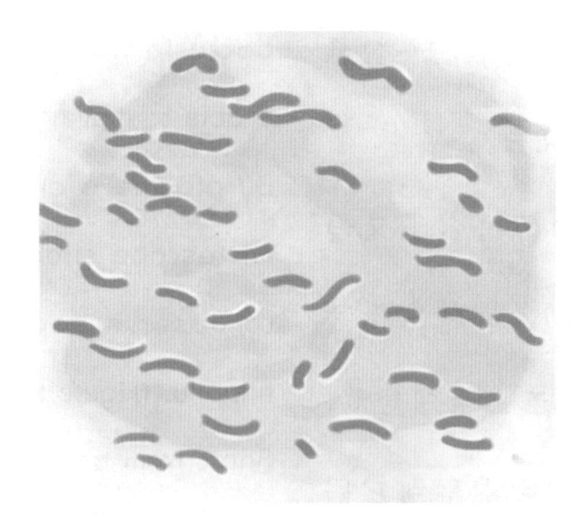

图 2.5　变形杆菌显微镜下鞭毛
Figure 2.5　Flagellum of Proteus under a microscope

图 2.6　变形杆菌显微镜下形态
Figure 2.6　The appearance of Proteus under a microscope

做才能把细菌和非细菌的物质分开呢？经过不断地探索，科学家们想到了给它们"穿衣服"的方法——染色。

早在1884年，丹麦医师Gram就创立了革兰氏染色法，成功地将细菌分成两个"帮派"——革兰氏阳性菌和革兰氏阴性菌。革兰氏阳性菌穿着厚厚的"外衣"（细胞壁），所以不容易被脱色而呈现紫色（图2.7）；而革兰氏阴性菌的"外衣"没那么厚很容易被脱色而重新染成红色（图2.8）。染完色的细菌在显微镜下原形毕露，有的胖、有的瘦、有的呈球形、有的呈杆状，于是科

exactly differentiate between bacteria and non-bacteria materials? Through constant research and discovery, scientists figured out a way to tell bacteria apart from other items, it's a way to put "clothes" on bacteria for easy recognition— dye.

In 1884, a Danish doctor named Gram established the dye method—the Gram stain, which successfully divided bacteria into two groups—Gram positive bacteria and Gram negative bacteria. Since Gram positive bacteria possess thick cell walls which detain dyes inside the cell, they appear to be purple after Gram stain (Figure 2.7). In contrast, cell walls of Gram negative bacteria are relatively thin so that they are

图 2.7　革兰氏染色图片——革兰氏阳性球菌
Figure 2.7　Gram stain–Gram positive bacteria

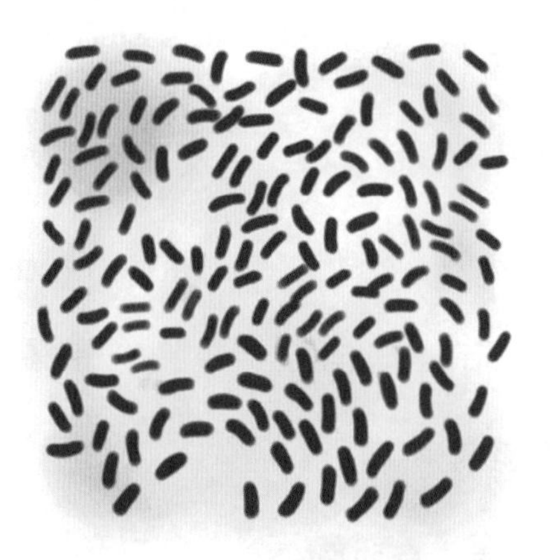

图 2.8　革兰氏染色图片——革兰氏阴性杆菌
Figure 2.8　Gram stain–Gram negative bacteria

学家们就根据“身材”对它们进行了分群——球菌、杆菌、球杆菌。

后来，科学家们发现革兰氏染色法不能很好地区分分枝杆菌和非分枝杆菌，所以又发明了抗酸染色。因为分枝杆菌的“外衣”细胞壁有很多脂质，这些特殊成分能够和染料牢牢地结合在一起，所以染成了红色，而其他非分枝杆菌则被染成了蓝色。此外，还有很多染色方法能帮助科学家们更好地研究细菌家族成员的特征。

像人体有一套完善的消化系统一样，细菌也有自己独特的“消化系统”——酶系统。因为不同菌种

easily decolorized and redyed to be red (Figure 2.8). After Gram stain, bacteria can be clearly observed under a microscope. According to their different shapes, scientists classify them into three groups—cocci, bacillus and coccobacillus.

Later, scientists found that it was difficult to distinguish mycobacteria from other bacteria through the Gram stain method.Then, the anti-acid dyeing method was invented. The mycobacteria's cytoderm has lots of lipid which can be combined with dye firmly, so they are dyed into the color red while others are blue. In addition, there are many dyeing methods to help scientists to better study and differentiate the characteristics of

所含的酶各不相同，这些酶的分解能力和消化后得到的产物千差万别，所以科学家们就想到了利用生物化学方法测定这些产物（即细菌的生化试验）来区别不同菌种。这些形形色色的生化试验主要包括碳水化合物的代谢试验、氨基酸和蛋白质的代谢试验、氮源碳源利用试验和酶类试验等。

某些细菌（如大肠杆菌、变形杆菌等）具有色氨酸酶，能够使含色氨酸的培养基变成红色(图 2.9)；有些细菌（如伤寒沙门菌）能消化培养基中的含硫氨基酸产生硫化氢，经过简单的化学反应在培养基

bacteria.

Like humans with a complete digestive system, bacteria have their own unique system as well—enzyme system. Scientists distinguish different bacteria through Biochemical methods because different bacteria contain different enzymes, which can be fermented into different discretion. These various biochemical experiments mainly include carbohydrate metabolism, amino acid and protein metabolism trial, nitrogen source and carbon source using experiments and enzymes tests, etc.

Some bacteria, such as *Escherichia coli* and *Proteus*, produce tryptophanase, which can turn tryptophan-containing medium into the color red (Figure

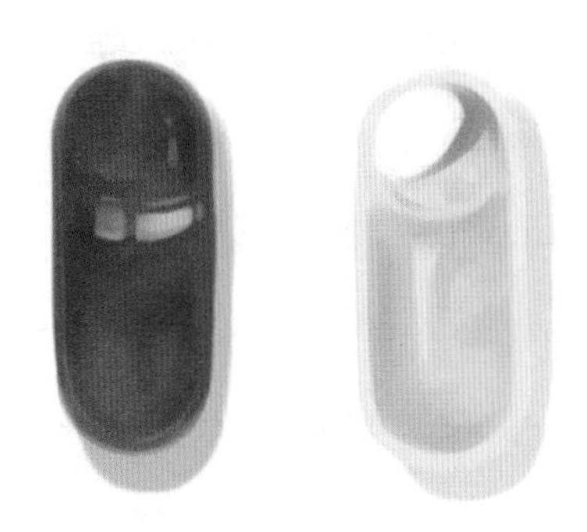

图 2.9 色氨酸阳性，呈红色
Figure 2.9 The positive reaction for Tryptophan—Red

上呈现黑色（图 2.10）；还有一些特殊的细菌，如产气荚膜梭菌，能

2.9). Others like *Salmonella typhi*, can digest sulfur—containing amino acid to produce sulfuretted hydrogen, which appears to be black in the medium through simple chemical reactions (Figure 2.10). Some special bacteria, such as *Clostridium perfringens*, can digest lactose into acidoid to curdle casein, following lots of gas which can scatter concretionary casein into honeycomb like figures (Figure 2.11). Another sort of

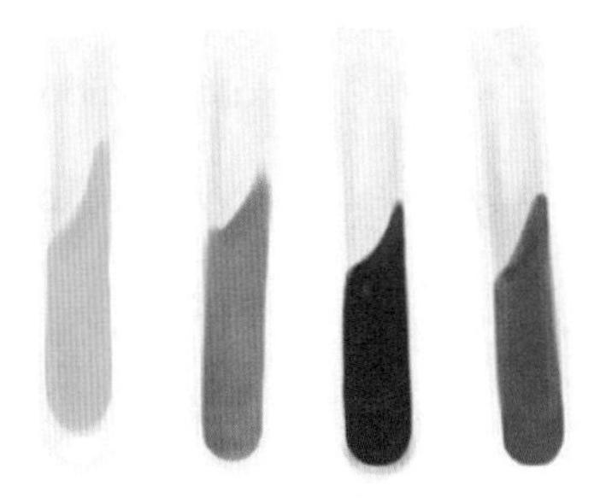

图 2.10 第三根管子为产硫化氢，培养基呈黑色
Figure 2.10 The third black mediumis hydrogen sulfide

消化乳糖产生酸性物质使酪蛋白凝固，同时产生大量气体，气势汹汹，将凝固的酪蛋白冲散成蜂窝一样（图2.11）；有些细菌，如无乳链球菌，能产生某些因子，加强金黄色葡萄球菌溶解红细胞的能力，因此在两菌交界处出现箭头一样的透明溶血区（图2.12）。

长久以来，生化反应是鉴定细菌的经典方法，但科学家们还是发现了很多“漏网之鱼”。生物的遗传物质如同一个巨大的“宝藏”，包含大量且大多都是很独特的信息，于是科学家们开始利用核酸分析技术对细菌进行“亲子

bacteria, like *Streptococcus agalactiae*, can release specific minarls to enhance the capacity of *Staphylococcus aureus* to dissolve erythrocyte, which result in a transparent hemolytic area near the border of two species (Figure 2.12).

For the longest time, biochemistry reactions were regarded as the most classical approach to identify bacteria. However, scientists discovered that this method had many flaws. The hereditary materials or an organism is like a huge treasure chest, containing large amounts of unique and precious information. Therefore, scientists started utilizing nucleic acid to conduct "paternity tests" for bacteria.

Deoxyribonucleic acid (DNA), carries lots of genetic information

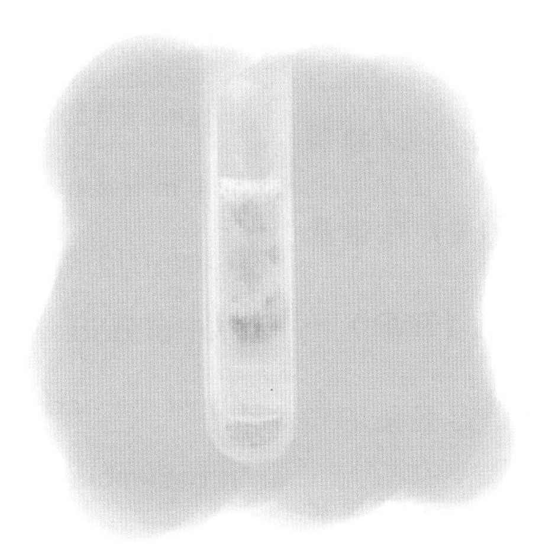

图 2.11 产气荚膜梭菌“汹涌发酵”现象
Figure 2.11 The phenomenon of clostridium Perfringensaerogenesis

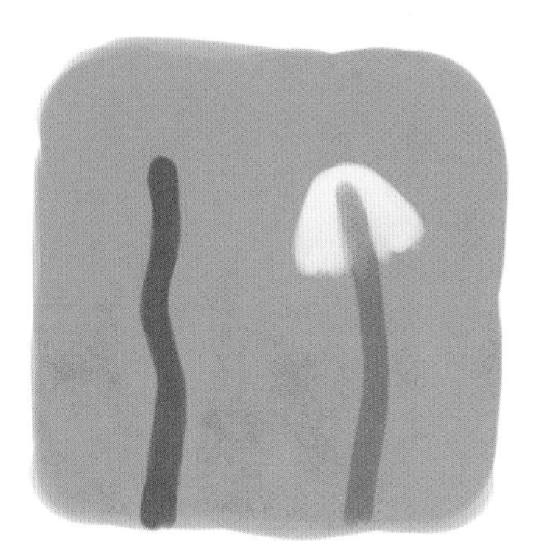

图 2.12 无乳链球菌加强溶血区——箭头
Figure 2.12 The arrow stands for Streptococcus agalactiae enhancing hemolysis

鉴定”。

脱氧核糖核酸（DNA）以独特的双螺旋结构（图2.13）携带庞大的信息，与核糖核酸（RNA）一起在生命活动中起着举足轻重的作用。科学家们研究发现，不同菌种

through the double helix structure (figure 2.13) and plays an important part in the activities of life alongside ribonucleic acid(RNA). The 16Sr RNA of bacteria, which is a part of ribosome and plays a dominant role in assembling protein, distinguishes each other among species. Similarly,

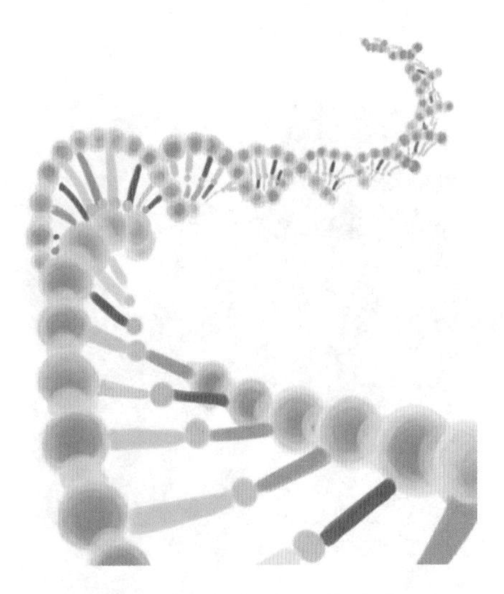

图 2.13 DNA 双螺旋结构
Figure 2.13 DNA double helix structure

的16Sr RNA（细菌体内组装蛋白质的机器——核糖体的成分之一）千差万别。类似地，真菌胞内的18Sr RNA也是差异巨大。借助聚合酶链反应，科学家们可以把极微量的细菌遗传物质成千上万倍扩增，然后通过对这些核酸物质进行测序解读信息达到鉴别的目的。

相比于测序法，基因芯片法更像是“大海捞针”，然而不要以为这是种笨方法哦。DNA是双链结构，两条链上的碱基互补配对，科学家们利用这一特性，合成能够捕获不同菌种特异性核酸序列的探针作为“吸铁石”。只要把这块吸铁石扔进

the 18Sr RNA of fungus cells varies from each other. The tiny genetic material of bacteria can be amplified by polymerase chain reaction (PCR), and then the genetic information can be read to identify species by sequencing.

Compared to sequencing, gene-chip is more like a needle-in-a-haystack, but don't think this idea is silly. DNA is double-chain structured with two complementary chains. Because of the above characteristics, scientists can compound specific nucleic acid sequence probe as "magnet" to capture different targets. "Needle" can find the "magnet" as soon as it is thrown into the ocean of nucleic acid. After a series of signal

核酸的海洋中，“针”很快就能被找到。通过一系列信号逐级放大和光电信号转化，机器能够准确地把菌种类别判定出来，既方便又快捷。

除了核酸方法鉴别菌种外，蛋白质指纹图谱法（以下简称质谱图法）也是一种极好的方法。质谱方法其实就是测量细菌核糖体蛋白的特异性图谱的方法，把得到的图谱和已知的图谱进行比较。这就像我们玩“连连看”，越像，分数就越高，从而来判断这个细菌是什么。这种方法是20世纪80年代新兴的一种方法，被认为是微生物鉴定中一项革命性的技术。

amplifier and photoelectric signal conversion processes, the machine can quickly sort the strains into different categories, this is both convenient and efficient.

In addition to the nucleic acid method to identify strains, protein fingerprint is also a wonderful method. The purpose of mass spectrometry method is to measure the specific profile of the bacterial ribosomal protein. To identify bacteria, we need to compare the obtained profile with the already known one. This was the emerging method back in the 1980s, it is considered to be a revolutionary technology in microbial identification.

借助核酸扩增、基因芯片、质谱技术等分子生物学技术，更多的细菌家族成员开始为我们所认识。

With the help of nucleic acid amplification, gene chip, mass spectrometry and other molecular biological techniques, more and more members of the bacteria kingdom will be known and understood.

第三章

有益菌和有害菌——细菌中的“警察”和“小偷”

Beneficial Bacteria and Harmful Bacteria—Police and Thieves in the Bacteria

1. 如何区分“好”细菌和“坏”细菌

形形色色、千奇百态的细菌无“微”不至，充斥在我们生活的所有空间。你知道细菌也像人类一样，有好人和坏人之分吗？

绝大多数细菌对人类、动物和植物都是有益的，有些还是必不可少的。很多腐生菌能够把动物、植物的尸体和排泄物等分解为水、二氧化碳和其他无机盐。无机盐则可作为肥料供给植物和农作物，植物又被我们人类和动物所食用。这就有了我们大自然的物质循环，有了

1. How to Distinguish "Good" and "Bad" Bacteria

We are surrounded by various kinds of bacteria in our daily life. Do you know that like humans, bacteria can also be divided into "good" and "bad"groups?

The majority of bacteria are beneficial to humans, animals and plants, some of them are indispensable to our environment. Many saprophytes can decompose the corpse of animals, plants and excrement into water, carbon dioxide,and other inorganic salts. Inorganic salts are used as fertilizers to plants and crops, which will be consumed by humans and animals. This makes up the substance

这个世界的生生不息（图 3.1）。

千万别小瞧了细菌这些“小东西”。绝大多数的抗生素都是它们的代谢产物。比如，青霉素就是从青霉菌中提炼出来的，链霉素是链霉菌所分泌的。细菌在污水、垃圾

图 3.1　细菌在自然界的物质循环中起着重要作用

Figure 3.1　Bacteria play an important role in the material circulation in nature

circulation of our ecosystem and enables us to survive and thrive(Figure 3.1).

Do not underestimate these tiny creatures. The majority of antibiotics are metabolites from bacteria. For example, penicillin is derived from *Penicillium*, streptomycin is secreted by *Streptomyces*. Bacteria also play an important role in garbage disposal, sewage treatment and the degradation of toxic substance.

Under normal conditions, bacteria that inhabit on the surface and the cavity of humans and animals are harmless. Moreover, some of them are also beneficial, such as *Bifidobacteria* colonized in the intestine. They can not only resist the invasion of other

的无害化处理、降解有毒物质等方面也起到了非常好的效果。

在正常情况下，寄居在人类和动物的体表和腔道中的细菌是无害的。有些还是有益的，如定居在肠道中的双歧杆菌，它们不仅能够抵抗其他外来细菌的入侵，还能为宿主提供必需的维生素和氨基酸等营养物质。

有些细菌（如肺炎链球菌）是口腔和鼻咽部正常居住的细菌，在正常情况下不会使人生病。但当身体抵抗力下降时，如在受凉、感冒后，肺炎链球菌会引起肺炎、脑膜炎等疾病。有些细菌（如大肠埃希

extrinsic bacteria, but also provide necessary vitamins, amino acids and other nutrients for their host.

Some bacteria (such as *Streptococcus pneumoniae*) are normal flora in oral cavity and nasopharynx. Under normal conditions, they will not cause illness. However, when the body's resistance is weakened, such as catching a cold, *Streptococcus pneumoniae* can cause pneumonia, meningitis and other diseases. Some bacteria (such as *Escherichia coli*) originally lived in the intestines of humans and animals,but when they moved to other parts of the host, they can cause infections. Such a well-behaved "good" bacteria, due to weakened resistance and translocation

菌）原本居住在人和动物的肠道中，但当它转移到了其他部位居住，就会引起人和动物其他部位感染。这类原本安分守己的“好”细菌，由于抵抗力下降、居住地改变等引起感染，导致体内的平衡被打乱时，它们也会乱了阵脚繁殖过量，甚至做起“坏事”来。这样的细菌，我们称其为条件致病菌。

有少数细菌能引起人类和动物、植物的病害，我们称之为致病菌。例如，霍乱弧菌、沙门菌、志贺菌、分枝杆菌、破伤风梭菌、炭疽芽孢杆菌等细菌，可分别引起人类的霍乱、伤寒、痢疾、结核、破

to different habitats, can over reproduce and disrupt the balance of our body. Ultimately, they will turn into "bad" bacteria. These types of bacteria are often called opportunistic pathogens.

A small number of bacteria can cause diseases of humans, animals and plants, and we call them pathogen. For example, *Vibrio cholerae* cause cholera, *Salmonella* cause typhoid, *Shigella* cause dysentery, *Mycobacteria* cause tuberculosis, *Clostridium tetani* cause tetanus, *Bacillus anthracis* cause anthrax. Some of the pathogens cause chicken cholera, duck plague, cattle anthrax and other diseases. The main pathogenic bacteria of plant diseases

伤风、炭疽等严重的疾病；或者使动物患鸡霍乱、鸭瘟、牛炭疽等疾病。植物病害主要病原细菌有假单胞杆菌属、黄单胞菌属、欧文氏菌属等。这些细菌可引起水稻白叶枯病，花生、辣椒、茄子青枯病，白菜软腐病等（图 3.2）。

are Pseudomonas spp., Xanthomonas spp., Erwinia spp. and etc. These bacteria can cause rice bacterial leaf blight, making peanut, pepper and eggplant bacterial wilt, and sometimes cause cabbage soft rot. (Figure 3.2)

图 3.2　水稻、白菜等的病害图
Figure 3.2　Disease of rice and cabbage

2. 环境中常见的“好”细菌和“坏”细菌

你们知道一克的土壤中有多少个细菌吗？答案是几亿至几百亿个。土壤中富含细菌爱吃的营养物质，而土壤的肥沃和生生不息也归功于不少爱护家园的细菌“卫士”。其中，有一些偏爱与植物互相扶持的固氮菌，如根瘤菌。它们独自生活的时候，以动植物残体为食，过着自给自足的“腐生生活”。而当土壤中有它们的“小伙伴”——豆科植物的生长时，根瘤菌会迅速向“小伙伴”的根部靠拢并进入其中。

2. "Good" and "Bad" Bacteria in the Environment

Do you know how many bacteria exist in one gram of soil? The answer is hundreds of millions to tens of billions of bacteria. Soil provides all kinds of nutrients that bacteria need. And as a return, bacteria help to retain the fertility of the soil. For example, the nitrogen-fixing bacteria, such as *Rhizobia* can absorb nutrition from the corps of animal and plant, they are able to live and thrive independently. And when his "partners"—leguminous plants appear nearby, *Rhizobia* quickly move towards and merge into the root of legumes. Due to the influence of *Rhizobia*, the roots of legumes

豆科植物的根部在根瘤菌的影响下迅速分裂膨大形成根瘤，慷慨地为根瘤菌提供家和粮食；同时根瘤菌又会卖力地从空气中吸收氮气，为“小伙伴”制作美味的“氮餐”，使其枝繁叶茂。根瘤菌生产出来的氮肥不光能满足“小伙伴”的需要，还会分出一些帮助“远亲近邻”，储存一部分养肥土壤（图3.3）。

而硝化细菌和反硝化细菌则致力于共同维护自然界中氨的平衡和氮的循环。还有一些腐生细菌对作物的残根败叶和有机肥料情有独钟。它们吃掉有机质，产出营养元素，不仅惠及作物，还形成腐殖质，

expand rapidly and supply food and shelter for *Rhizobia*. At the same time, the *Rhizobia* absorb nitrogen from the air and change it into delicious food, which will help the growth of its partner. The nitrogen fertilizer produced by *Rhizobia* not only meets the needs of legumes, but also fertilizes the soil surrounding the root (Figure 3.3).

The nitrifying bacteria and denitrifying bacteria together contribute to maintain the balance of ammonia and the nitrogen cycle in nature. Some saprophytic bacteria have preference for decaying roots and leaves. They degrade organic material and produce nutrients which greatly benefit the crops and increase

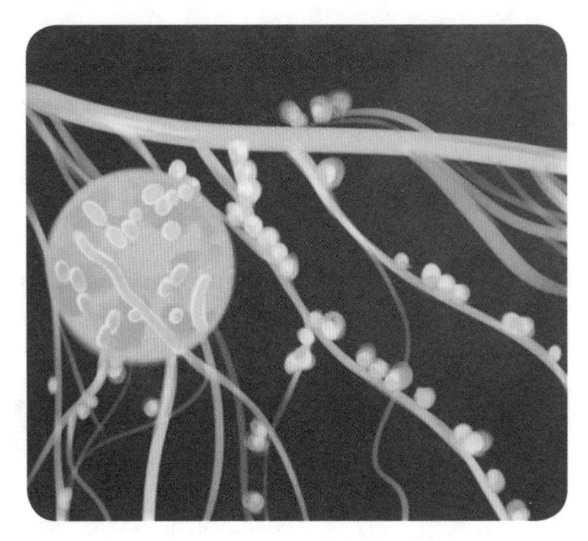

图 3.3 豆科植物根部的根瘤菌
Figure 3.3 Rhizobium in the root of legume

改善土壤的结构和耕性。不仅如此，产生抗生素细菌为了保护“同伴邻里”，会分泌抗生素以大大削弱病原菌的战斗力。土壤微生物甚至共同合作把土壤中残留的有机农药、城市污物和工厂废弃物等降解

tilth. Some bacteria have the ability to produce antibiotics which weaken the activity of pathogens and protect their "neighbors". Soil microorganisms can even work together to convert the residual pesticide in soil, urban sewage and factory waste into less harmful or harmless substances, which

成低毒乃至无害的物质，极大地造福人类。

细菌在水体的自净作用中也功不可没。由于光的穿透性、水的温度和水中含氧浓度的不同，不同的微生物群体占据着不同层次的水体。在水产养殖生态环境中，光合细菌、硝化细菌和芽孢杆菌等以大量残余的饵料、水生动物排泄物、动植物残体和有害气体为食，净化水质并为浮游植物提供营养促进其光合作用产氧，使水产动植物越发健康茁壮地生长（图3.4）。

当然，土壤和水体中也有一些自私自利甚至伤及他人的细菌。有

benefits humankind greatly.

Bacteria also contribute a lot to the capacity of self-purification in the water. Different microbial populations occupy various level of water due to differences in light penetration, water temperature and oxygen concentration in the water. In the ecological environment of aquaculture, *Photosynthetic bacteria*, *Nitrifying bacteria* and *Bacillus* consumes a large quantity of residual foods, aquatic excrements, animal/plant debris and harmful gases,contributing to water purification and providing nutrients for phytoplankton to thrive and perform photosynthesis (Figure 3.4).

Of course, soil and water also contain certain bacteria that

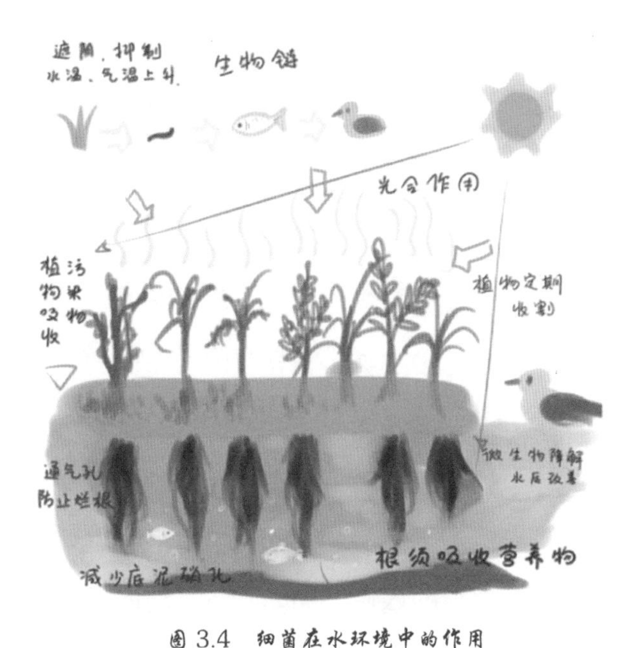

图 3.4 细菌在水环境中的作用

Figure 3.4 The role of bacteria in the water environment

益菌和有害菌是一个矛盾的存在。它们相生相克，此消彼长，通过环境与营养的竞争、彼此代谢物的相互影响而发生极其复杂的相互作用。而这场斗争中的胜者将会成群

are selfish and harmful to others. Beneficial and harmful bacteria are always battling and fighting against each other. They compete within an environment for food and nutrients, and their metabolite also interacts

结队地生长壮大并成为优势菌群，决定疾病的发生与否。

植物土传病原菌，如棉花黄萎病菌、番茄青枯病菌等，一旦缠上作物，就能迅速地致植物枯萎、死亡。若农田在灌溉或底泥施肥过程中遭受未经处理的人畜粪便、生活污水、垃圾甚至含有病原体的医院污水和工业废水等的污染，就会把大量致病性细菌带入土壤，如沙门菌、志贺菌等，引起人体感染各种细菌。

尽管大部分细菌会因空气缺乏营养和适宜的温度而难以为继，甚至会因阳光的照射和干燥而死

in complicated way. The winner of the battle will become the dominant bacteria group, and they will determine the occurrence of diseases.

Plant soil-borne pathogens, such as *Verticillium dahliae*, *Ralstorinia solanacearum* and etc., can quickly cause crops to wither and die. If the farmland is polluted by the untreated human and animal waste, sewage, garbage, hospital sewage, and industrial waste water, these medium will bring disease causing pathogens (such as *Salmonella*, *Shigella*, etc.) into soil and the process of irrigation and sediment fertilization.

Although most bacteria cannot survive in the air due to lack of nutrients and improper temperature,

去，但一些生命力顽强的细菌和真菌，如芽孢杆菌和真菌孢子，依然不畏惧恶劣的环境而坚强地活着。某些狡猾的病原菌，如结核杆菌、产气荚膜梭菌、葡萄球菌等，会乘着病人的飞沫、皮屑、痰液等飘浮在空气中，伺机侵害缺乏防备的弱势群体。

细菌还有一个讨喜的重要职业角色——美食名庖。在酿酒业，美酒独特的清香和醉人的口感可是需要乳酸杆菌、醋酸菌、丁酸菌或枯草芽孢杆菌等细菌的加工。

and some even may die due to sunlight and drying, nevertheless, some tenacious bacteria and fungi, such as bacillus and fungi spores, are very resilient. On the other hand, some clever pathogens, such as *Mycobacterium tuberculosis*, *Clostridium perfringens, Staphylococcus* spp., they often hangs around in the patient's cough aerosol, dander, sputum and etc., waiting to attack and compromise those with a weak immune system.

Bacteria also hold another important profession—a gourmet chef. Winery produces unique wines with marvelous tastes, and the production is assisted with the help of *Lactobacillus, Ocetobacter, Butyric acid bacteria,*

然而，过犹不及同样适用于它们，一旦它们用力过猛就会毁了美酒的醇香。清香甘醇的牛奶在乳酸杆菌的魔法下摇身一变为酸甜醇厚的酸奶。还有清脆酸甜的泡菜也是肠膜明串珠菌、植物乳杆菌、乳球菌等引以为豪的佳作（图3.5）。

bacillus subtilis and other bacteria.

However, the overuse of bacteria will destroy the taste of the wine. With the magic of *Lactobacillus*, delicious milk can be turned into delicious yogurt. Crisp, sweet and sour kimchi is the masterpiece of *Leuconostoc Mesenteroides*, *Lactobacillus Plantarum* and *Lactococcus* (Figure 3.5).

图 3.5 营养又美味的酸奶和泡菜

Figure 3.5 Nutrition and delicious yogurt and pickles

3. 动物体内常见的“好”细菌和“坏”细菌

3. "Good" and "Bad" Bacteria in Animals

动物体内常见的细菌也有好有坏，我们先来看看几种“好”细菌吧。“好”细菌就像在动物体内生活的小卫士，在不同的岗位上尽职尽责地工作，促进消化吸收、提高免疫力、保持体内微生物平衡，它们都贡献着自己的力量。

Bacteria in animals can also be either beneficial or harmful. First, let's take a look at some "good" types bacteria. "Good" bacteria are just like guards living in animals, they work diligently in different positions to help the animal to digest and absorb nutrients, as well as improving the immunity system and maintaining the balance of microorganisms in the body.

乳酸菌是一种存在于各种动物体内的益生菌，它在动物体内能够调节胃肠道里的正常菌群，让生活在里面的各种细菌保持平衡，帮助动物消化食物，同时制造营养物质，

Lactobacilli are a group of probiotics which exist in various animals. They can regulate and maintain the balance of the normal microflora in the gastrointestinal tract, promote

还可以刺激组织发育，控制内毒素，抑制肠道内腐败菌生长繁殖和腐败产物的产生，对动物身体里的许多生理过程都产生作用。

芽孢杆菌也是一种常见的有益菌，广泛存在于土壤、水、尘粒和空气中，可以通过食物被吃进肚子，从而进入动物的体内。进入体内的芽孢杆菌可与宿主之间形成一种奇妙的共生关系，就像暂时寄宿在动物胃肠道中的房客。但这群房客可比那些致病细菌有礼貌得多。宿主动物"包吃包住"，它们也会以各种方式回报宿主。

比如，益生芽孢杆菌能在胃肠

digestion, create nutrients, stimulate tissue development. They can even reduce the toxicity of endotoxin, inhibit the growth of spoilage bacteria and the production of bad excretions in the intestine.

Bacilli are another common group of beneficial bacteria, which widely exists in soil, water, dust and air. They can enter into stomach through the consumption of food. *Bacilli* can form a symbiotic relationship with the host, like a temporary tenant in the animal gastrointestinal tract. These tenants are more polite than pathogenic bacteria. The hosts provide shelter and food for *Bacilli*, in return, *Bacilli* pay back the host in a variety of ways.

For example, those tenants in

道代谢产生一些肽类抗菌物质，并通过"竞争排斥"抑制其他外来病原微生物在胃肠道内的生长，还能刺激免疫系统，让动物的抵抗力更强。另外，芽孢杆菌体外都具有较强的蛋白酶、脂肪酶和淀粉酶活性，这些消化酶类能够促进动物对营养物质的吸收，提高饲料的转化效率。

1899年，法国学者Tissier从吃母乳的婴儿粪便中分离出来一种厌氧革兰氏阳性杆菌。因为它们的末端常常分叉，所以被叫做双歧杆菌。双歧杆菌是益生菌中的一大类，动物或人出生3~4天后肠道内就会出现，随着年龄的增加，双歧杆菌

the gastrointestinal tract can produce some antimicrobial peptides which can inhibit the growth of ectogenic pathogenic microorganisms through competitive exclusion. Moreover, they can also stimulate the immune system and improve the body's level of resilience. Additionally, Bacillus contains strong extracellular protease, lipase and amylase that are highly active. These digestive enzymes can promote the absorption of nutrients in animals and improve animal feed conversion rate.

In 1899, a French scholar named Tissier isolated a group of anaerobic Gram-positive strains from feces of a breast-fed baby. These tiny creatures were then named *Bifidobacteria*

的数量也会逐渐减少。在动物肠道中，双歧杆菌可以帮助动物消化它们本来不能消化的食物，促进维生素的吸收和利用。同时，双歧杆菌通过提高动物免疫力，还可发挥抗感染及抗肿瘤作用。双歧杆菌对生存环境的要求十分苛刻。它们数量的减少甚至消失是机体“不健康”状态的标志，因此它们还被誉为健康的“晴雨表”。

有益的细菌还有许多，但危害动物健康的“坏”细菌也有不少。

葡萄球菌是一类自然界中广泛分布的革兰氏阳性菌。它在显微镜下呈圆形，聚在一起像葡萄

because of their split ends. The *Bifidobacteria* belong to probiotics, they appear in the guts of humans and animals 3~4 days after birth. The number of these bacteria gradually reduces with age. The *Bifidobacteria* can help animals to digest food which they cannot digest originally in their guts and improve the absorption and utilization of vitamins. At the same time, the *Bifidobacteria* can improve the immunity of animals, thus play an important role in anti-infection and tumor-suppression. The *Bifidobacteria* are highly demanding to their living environment. A decrease in *Bifidobacteria* numbers indicates our bodies are unhealthy, they are known as the "barometer" of health.

串一样，因此被命名为葡萄球菌。葡萄球菌分很多种，其中金黄色葡萄球菌是葡萄球菌中一类主要的病原菌。金黄色葡萄球菌在自然界中无处不在。因此动物饲料很容易被污染，动物吃到被污染的饲料会腹泻。同时，由于动物的皮肤和黏膜暴露在空气中，或间接接触空气。一旦皮肤或黏膜受到损伤，也很容易让金色葡萄球菌趁虚而入，从而导致发炎。

沙门菌是一种没有芽胞的直杆菌，全身长了很多可以帮助它们运动的鞭毛（图3.6）。沙门菌在很多动物体内都有，能引起幼年和青年

There are many beneficial bacteria, but there are also plenty of bad bacteria that are harmful to animals.

Staphylococci are a type of Gram-positive bacteria commonly found in nature. Under the microscope, they appear to be round in shaped grouped together like grapes, that's how they got their Chinese name for *Staphylococci*. There are many species among *Staphylococci*, and *Staphylococcus aureus* is one of the major pathogens. They are every where in nature, so animal feed can be polluted rather easily by then. Animals can get diarrhea once they eat the contaminated feed. Additionally, animals may be easily infected by *Staphylococcus aureus* if they are hurt

图 3.6　显微镜下的沙门氏菌
Figure 3.6　Salmonella under microscope

动物的败血症、胃肠炎及其他组织局部炎症。科学家们通常根据沙氏菌对宿主的挑剔情况，把它们分成三个帮派：专嗜沙门菌，只对某种动物产生特定疾病，比如仅引起鸡和火鸡发病的鸡白痢和鸡伤寒沙门菌；偏嗜沙门氏菌，比较“挑食”，

or wounded,and this may lead to inflammations when the animals' skin and mucosa are exposed in the air, or indirectly contacting with air.

Salmonella are a group of straight bacilli without spores, and covered with flagella to move around with (Figure 3.6). *Salmonella* exist in many animals, and can lead to septicemia, gastroenteritis and other local inflammation among young animals. In terms of different hosts, *Salmonella* are divided into three main groups. Those that only infect specific animals are called the obligate *Salmonella*, includes *Salmonella pullorum* and *Salmonella gallinarum*, which only infect chickens and turkeys.Pan-salmonella are not picky and they can

特别喜欢感染某一种动物，但其他动物有时也可以感染；泛嗜沙门氏菌最不挑剔，在许多动物体内都能存活，能引起人和各种动物的沙门菌病，所以危害也是最大的。在细菌源性食物中毒的“罪魁祸首”排名中，沙门菌位列榜首。大多病例都是由于吃到沙门菌感染的肉、蛋引起的，所以为了我们的健康，平时不要生吃蛋和肉哦！

布氏杆菌是一群危险的家伙，对人类和许多动物都有很大危害，因此备受重视。布氏杆菌分很多种，都有各自喜欢感染的动物，在中国，最常见的易感动物是羊。通常，细

survive in many animals, thus causing salmonellosis to be the greatest threat to public health. Among the pathogens that causes food poisoning, *Salmonella* rank at the top of the list. Most of the cases are caused by eggs and meat that are contaminated by *Salmonella*, so for our safety, it is better not to eat raw eggs and meat.

The *Brucella* are a dangerous bunch. They deserve more attention because of the great harm they pose to human and many animals. They are divided into many species and every species have specific animals they target. In China, sheep are the most vulnerable type of animal. Usually, bacteria are killed by phagocyte after entering the body, and then

菌进入体内后，会被体内的吞噬细胞吃掉，然后被吞噬细胞内的溶菌酶等物质杀死。没被杀死的会被送到一个叫淋巴结的地方，被许多吞噬细胞一起消灭。布氏杆菌非常狡猾，不但不会死，还能寄生在吞噬细胞里面，甚至进一步攻击淋巴结。其实，它们最主要的攻击对象是动物的生殖系统。因此，怀孕的动物对布氏杆菌格外敏感，感染后常引起流产。布氏杆菌能通过气溶胶传播，是潜在的生物武器。人和带菌的羊接触感染后，也容易引起不孕等症状。因此，大家和小动物——尤其是和羊一起玩耍的时候，千万

most of them are killed by lysozyme in phagocyte. The survivors are transported into the lymph node, and then eliminated by phagocytes. However, Brucella are too clever to be killed. They parasitize in phagocytes even attacking the lymph node. Actually, the foremost target is the reproductive system of animals, so the pregnant animals are extraordinarily susceptible to *Brucella*, where it can cause a miscarriage. The *Brucella* can spread through aerosol, therefore they are potential chemical and biological weapons. *Brucella* may cause infertility to humans when they make contact with sheep carriers. It is important to be extra careful when playing and interacting with animals, especially

要多加小心（图3.7）。

动物体内还存在很多条件致病菌，如大肠杆菌，是人和恒温动物肠道中正常的菌群之一。在大多数情况下，大肠杆菌和宿主之间是互惠互利的共生关系。在宿主抵抗力

sheep. (figure 3.7).

There are also many opportunistic pathogens living in animals, such as *Escherichia coli*, which are normal florasin guts of human and some animals. In most cases, *Escherichia coli* and human are in symbiosis obtaining mutual benefits. Nevertheless, they

图 3.7 人和羊的亲密接触可能会感染布氏杆菌
Figure 3.7 The close contact with sheep leading to brucellosis infection in human

下降时，大肠杆菌却很容易从肠道跑到其他器官去，引起其他部位发病。当然，这个“家族”中也有几个“坏小子”，天生就爱搞破坏。例如，产肠毒素大肠杆菌会引起小动物腹泻。刚出生的小动物感染后会因剧烈水样腹泻而导致脱水死亡。因此在动物养殖业中，也需注意预防大肠杆菌。

can be easily translocated to other organs and cause diseases once the host's immune system is weakens. Of course, there are some "bad guys" in this family and those guys are rather disruptive. For example, enterotoxigenic *Escherichia coli* may infect newborn animals causing acute diarrhea, and they may die due to dehydration as a result of diarrhea. Therefore, prevention of *Escherichia coli* is important and also necessary in the animal feeding and breeding industry.

4. 人体中常见的“好”细菌和“坏”细菌

4. "Good" and "Bad" Bacteria in Human Beings

我们每天都生活在无数细菌的包围下，大多数细菌对人体是无害

Every day, we are surrounded by

的。它们寄居在人体的皮肤、肠道、口腔、鼻腔等腔隙中，帮助我们阻挡各种外来有害细菌的侵入，还能提供人体所需的营养物质等。它们的“喜怒哀乐”，影响着人类身体的方方面面。

科学家们发现，人体皮肤是细菌的“理想家园”。健康人体的皮肤上共居住了大约1000种细菌。其中，葡萄球菌、类白喉杆菌、铜绿假单胞菌等是主要的“常住居民”。它们构成了我们人体第一道天然的屏障，具有抵御外界病原体入侵的作用。

millions of bacteria, but most of them are harmless to our body. They live on our skin, within the bowel, oral cavity, and nasal cavity. Good bacteria can also protect us against invasion of other bacteria, and provide us with various essential nutrients. Their activities affect all aspects of our lives and health.

Scientists have discovered that the human skin is the most ideal habitat for bacteria. The healthy skin of humans may carry up to 1000 different types of bacteria, including *Staphglococcus aureus, Corynebacterium diphtheriae, Pseudomonas aeruginosa*. Bacteria form the first natural barrier for the human body against potentially harmful pathogens.

人体的胃肠道就像细菌的“动物园”。据估计，一个健康的成年人的胃肠道中寄居约有10^{14}个细菌，种类高达1000多种，它们有个响亮的名字，叫肠道菌群。它们是人体内最庞大的正常菌群，构成了一个巨大而复杂的生态系统。肠道菌群不仅是肠道忠实的“守卫兵”，抵抗各种外来敌人（致病菌）的入侵，而且也是称职的“营养医师”，为人体消化食物提供所需的营养，增强机体的免疫力。

The gastrointestinal tract of the human is like a “zoo” for bacteria. An estimate of 10^{14} bacteria and over 1000 species dwell in the gastrointestinal tract of a healthy adult. They have well known name called intestinal flora. They are the largest normal flora in the human body, constituting a large and complex ecosystem. Gut flora is not only the faithful “guards” against all kinds of invasion of extraneous enemies (pathogen), but also outstanding “dietician”. They help the body digest foods and provide essential nutrients for human while enhancing the hosts' immune system.

这些“小家伙”可不是与生俱来的。刚出生的婴儿肠道内是无菌的，但出生后几个小时就有细菌的进入。伴随着各种食物的进入，更多的细菌来到肠道这个“安乐园”定居。肠道菌群可分为常住菌（正常菌群）和过路菌这两种。前者是保持肠道安稳的主要群体，后者通过我们的进食进入肠道。常住菌有拟杆菌、乳酸杆菌、双歧杆菌、大肠杆菌和肠球菌等。过路菌有金黄色葡萄球菌、铜绿假单胞菌、产气肠杆菌、变形杆菌等。当然，在人体中也存在条件致病菌。临床常见的葡萄球菌引起的肠炎，就是肠道

These “little guys” are not inherent. Newborn baby’s intestinal tract is sterile, but soon after several hours after being born, bacteria will move in. With various kinds of food entering into intestinal tract, various bacteria settle in this “paradise”. Intestinal flora can be divided into resident bacteria (normal flora) and passing bacteria. The former is the major group of bacteria to maintain the balance of intestinal flora, and the latter passes through the intestines alongside food. Resident bacteria include Bacteroides, *Lactobacillus*, *Bifidobacterium*, *Escherichia coli* and *Enterococcus*, while the passing bacteria include *Staphylococcus* aureus, *Pseudomonas* aeruginosa,

中的“过路菌”葡萄球菌打败“常住菌”的一个典型的例子。

前面我们已经讲过动物体内的双歧杆菌。它在人体中也是“好细菌”的代表，素有“肠道清道夫”的美称。它可以合成B族维生素（包括B1、B2、B6、B12）、维生素K、烟酸和氨基酸等人体必需的营养物质，对人体的健康大大有益。这些营养物质到底有什么用呢？

举个例子，如果我们身体缺少B族维生素和氨基酸，就会导致我们漂亮的头发脱落、发黄、分叉。除此之外，双歧杆菌还能抑制肠道

Bacterium aerogenes, *Proteus*, and etc. Of course, there are also opportunistic pathogens in our body. Among them, enteritis caused by *Staphylococcus* aureus is the most common type of bacteria, typical example of "passing bacteria" beating "resident bacteria" in the human body.

We have already mentioned the *Bifidobacterium* in animals, and they are also the representative of "good bacteria" in the human body, known as "intestinal scavenger". They cansynthesize vitamin B (B1, B2, B6, and B12), Vitamin K, niacin, amino acids and other essential nutrients which are greatly beneficial to the health of human. What are the uses of these nutrients?

内过路菌的繁殖，减少某些毒素和致癌物质的产生。酸奶中就含有双歧杆菌，所以平常饮食中我们应该多喝酸奶，来补充双歧杆菌和乳酸杆菌等益生菌。

我们从小就接种过各种各样的疫苗。不管是吃的糖丸还是打的针，这些疫苗都是用来保护我们免受坏细菌引起的传染病。例如，百白破疫苗是用来预防百日咳杆菌、白喉棒状杆菌、破伤风梭状芽孢杆菌引起的疾病，卡介苗则是用来预防结核分枝杆菌引起的结核病。除了这些可以通过疫苗打败的“坏细菌”外，还有哪些“坏细菌”呢？炭疽

For example, lack of B vitamins and amino acid will cause our beautiful hair to experience hair loss, loss of color, and split ends. *Bifidobacterium* can also inhibit the regeneration of passing bacteria in intestinal tract and reduce the number of toxins and carcinogens in our bodies. Yoghurt contains *Bifidobacterium*, so we should drink yoghurt in our daily diet as supplement for *Bifidobacterium*, *Lactobacillus* and other probiotics.

We have received all kinds of vaccines since we were young. Whether it's through needle injection or sugar pills, these vaccinations are all used to protect us from harmful infectious diseases caused by bacteria. For example, the DPT vaccine

杆菌、霍乱弧菌、鼠疫杆菌、肉毒杆菌等，这些坏细菌引起的疾病曾经肆虐，夺走了很多宝贵的生命，对社会公共卫生和经济的发展造成极大影响（图 3.8）。

细菌无处不在。知道了“好细菌”的特点，才能更好地利用它

图 3.8 烈性传染病隔离区
Figure 3.8 Fulminating infectious diseases quarantine areas

is used to prevent the diseases that are caused by *Bordetella pertussis*, *Corynebacterium diphtheriae*, and *Tetanus bacillus*. The BCG vaccine is used for the prevention of tuberculosis (TB) caused by *Mycobacterium tuberculosis*. Besides the "bad bacteria" that can be beaten by vaccines, are there any other "bad bacteria" around? For example, *Bacillus anthracis, Vibrio cholera, Plague bacillus, Botulinum* which can cause diseases, and they took a lot of vivid lives away and caused adverse effect on the development of our social economy (Figure 3.8).

Bacteria are everywhere. Only when we have a better understanding of the characteristics of the good

们。仔细研究“坏细菌”的习性，不断探索“坏细菌”的另外一面，才能争取把“坏细菌”变成“好细菌”。

bacteria, can we take better advantage and utilize them for good. On the other hand, exploring deeper into harmful bacteria's behavior can aver tharmful bacteria into beneficial bactiera.

第四章

消灭细菌的方法

Ways to Kill Bacteria

1. 环境消毒和灭菌方法

在我们所在的世界里，平等地生存着各种各样的生物。窗外会唱歌的小鸟和会开花的树，是我们看得见的邻居，而无处不在的细菌，则是我们看不见的邻居。大家已经知道，细菌有好有坏，好的细菌我们要多多利用；而对于坏细菌，当然要找到它们的弱点，把它们通通消灭掉（图 4.1）。

大家可能听说过“消毒”和“灭菌”这两个词。有谁知道它们的区别呢？其实这两个词，都是指消灭细菌病毒等微生物。不同的是，消

1. Disinfection and Sterilization of Environment

In the world we live in, various creatures live among each other equally. Singing birds and flowering trees outside the windows are our nearby neighbors that we can observe, while the bacteria all around us are the invisible ones. As we know, there are both good and harmful bacteria living among us, we should make good use of beneficial bacteria, As for the harmful ones, we should certainly search for their weaknesses to eliminate them all (Figure 4.1).

Maybe we all have heard the two words "disinfection" and "sterilization", but who can tell the difference be-

图 4.1 各种消毒灭菌方法
Figure 4.1 The summary of disinfection methods

毒要求比较低，只要消灭对我们不利的病原微生物就好，而灭菌则是一视同仁，所有的细菌都被我们通通消灭掉。这两种方法在不同的情况下，有不同的用处。

我们生活的环境存在着许许多

tween these two? In fact, the two words are both meant to represent methods to kill microorganisms such as bacteria and viruses. The difference is the requirement for disinfection is lower, it can eliminate the harmful microorganisms, but sterilization is to kill all the bacteria. The two methods are applied to different situations.

The environment where we live in exists many bacteria. Even in a clean environment, there can be about 4000 bacteria per cubic meter of air, while air within a confined space or stagnant air contains even more bacteria. Floating bacteria in the air can easily spread diseases if it is not under control. Like humans, bacteria also enjoy a warm and comfortable environment to live

多的细菌，即使是在洁净的环境中，每立方米空气中大约有4000个细菌，而空气不流通的房间则更加严重。细菌会随着空气飘来飘去，如果不加以控制，很容易传播疾病。细菌像我们一样，喜欢在温暖而舒适的环境里生存。冬天天气比较冷，许多人不喜欢开窗，细菌在室内越长越多，这样非常不利于我们的身体健康。经常开窗通风，虽然不能把细菌杀死，但能在短时间内让室内外空气交换，减少室内的致病菌，这样也就降低了我们生病的可能性。

天气晴朗的时候，人们常会把

in. The weather gets cold during the winter, and most people don't like to open the windows, consequently, bacteria propagate easily within the enclosed space and it may be bad for our health. Although bacteria cannot be killed through opening the window, the exchange of air in a short frame of time can decrease the number of pathogenic bacteria indoors, thereby reducing the possibility of causing sickness to us.

During a nice and sunny day, people usually take out their blankets and sheets to put them out under the sun. Actually, this is a common method of sterilization. The sunlight creates a hot and dry condition that is terribly uncomfortable for bacteria, the ultraviolet ray is lethal to them as well,

被子拿出去晒一晒。其实，晒太阳也是我们平时经常用到的一种杀菌方法。太阳的干热会让细菌非常难受，阳光中的紫外线对它们来说更为致命，所以阳光越强，照射时间越长，杀菌的效果就越好。只是阳光中的紫外线经过空气的层层阻挡，照到地面上时已经很弱了。透过玻璃效果更是大打折扣。通常，在日光下直接曝晒4~6个小时，并且经常翻动，就能达到较好的效果。

生活中能用到的消毒灭菌方法还有很多，也各有利弊。比如，高压煮沸比较适合给毛巾、餐具等不

so the stronger the sunshine and the longer the time spent under the sun, the more effective the sterilization will be. However, the sunlight's ultraviolet ray decrease in strength as it passes through the multiple layers of the atmosphere, so by the time it reaches the ground, it is already very weak, and sunlight passing through glass is even weaker. Usually, directly laying things under the sunlight for 4~6 hours and constantly turning it over is the best method for sterilization.

There are many other useful sterilization methods that we can use in everyday life, and each has its own advantages and downsides. For example, the high pressure boiling method is suitable for high temperature

怕湿又耐高温的物品消毒；燃烧法简单迅速，只是操作起来有点危险，大家可千万不要轻易尝试。还有化学方法也很方便，对于墙壁、地面、桌椅之类的地方，用消毒液擦拭或浸泡就能消灭其表面的细菌；用食醋熏蒸可以对室内的空气消毒，是个预防流感的好办法。

现在也有很多空气净化机或者自净器。它们是新科技的代表，可以像警察一样，抓住空气中的坏蛋致病微生物，给我们创造一个清洁舒适的生活环境。

and water resistant products like towel and tableware, while the combustion method is simple and quick but a bit dangerous to operate, so we shouldn't try it under normal circumstances. In addition, chemical disinfection is also very convenient, it can kill bacteria living on the surface of tables, chairs and the ground. Vinegar fumigating is not only a good way to sterilize indoor air but it also helps prevent flu and sickness.

Now there are many kinds of air purifier or machine cleaners. They are the representative of new technology and innovation. They can catch the harmful microorganisms in the air, just like a policeman, and create a comfortable living environment for us.

On the other hand, we cannot

除此之外，“饭前便后要勤洗手”，这些被家长和老师唠叨了无数遍的话可不能小瞧。良好的卫生习惯对保护我们免受病原微生物侵扰都是十分有帮助的。正确的洗手方法可以消除手上大部分细菌，让我们“左手右手一个慢动作”，赶快学起来吧（图4.2）！

ignore these tips, "washing hands before meals and after using the toilet". These phrases have been emphasized countless times by parents and teachers, and these good hygiene practices are very helpful to protect us from pathogenic microorganisms. The proper hand-washing techniques can eliminate most bacteria living on our hands, so let's learn it right now (Figure 4.2!).

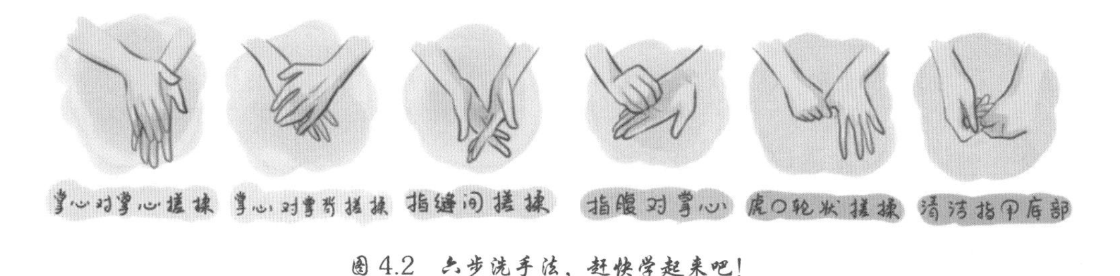

图4.2 六步洗手法，赶快学起来吧！
Figure 4.2 The correct way to wash your hands, let' hurry to learn!

2. 动物和人体抗菌方法

从出生开始，我们就生活在被细菌包围的环境中。即使平时再小心，也总有些狡猾的病原菌能够趁虚而入，悄悄地进入我们的身体。这样的情况几乎每天都在发生。可并不是每次细菌入侵都会让我们生病，这又是为什么呢？其实这一切，都是免疫系统的功劳。

在与细菌不断战斗的漫长岁月中，我们的身体逐渐进化出了一套完整的防御机制。而这套完成自我保护功能的系统，就叫做免疫系统。

免疫系统有几个特点，首先，

2. Anti-microbial Treatments for Animals and Humans

Ever since the day of birth, we have been living in an environment surrounded by bacteria. No matter how careful we are, certain clever pathogens will always figure out a way to enter our bodies. That occurs on a daily basis, but not every invasion of bacteria can cause sickness. Why is that? This is all the doings of our immune system.

The many years of fighting against bacteria and harmful microorganisms, our body has gradually evolved out a complete set of defense mechanisms against attacks. This complete automatic protection system is called

它的识别能力非常强，可以分辨出对方是自身的好细胞还是外来的敌人。如果是敌人，就毫不留情地攻击，如果是自身的细胞受损了，或是老得不能再继续为身体工作，免疫系统一样会把它们清理掉。如果这个功能坏掉，免疫系统变得好坏不分，身体就会出问题，自身免疫病就是这样产生的。

免疫系统还有个特点就是"记性"好，遇到见过的"坏人"，可以更快地做出反应，大量产生一种专门对付这种"坏人"的武器（也就是抗体），从而更有效率地把敌人消灭掉。根据这个特点，我们发明

the immune system.

The immune system has several characteristics. First of all, the immune system has a quick eye, it can distinguish foreign enemies from autologous cells. It will mercilessly attack invasive enemies, and also clean damaged or old cells who can't work for the body any longer. But if this function is broken, the immune system would fail to tell the good cells and the bad cell apart, which then causes human diseases named autoimmune diseases.

Another key feature of our immune system is that it has a good memory. When encountering harmful microorganisms who have ever met before, it can react faster and produce

了疫苗。疫苗通常是把很厉害的病原微生物的毒力减弱或杀死之后做成的。它们进入我们体内后，不会引起严重的疾病，但会刺激到免疫系统，让免疫系统经历一场小规模的实战演习。

这样，当环境中致病力很强的同一种病原菌再次潜进我们身体的时候，就会发现自己面对的是一支训练有素的"小分队"。它们快速地产生大量让自己害怕的抗体，病原菌就只好落荒而逃了。所以，下次打疫苗的时候可别再哭鼻子啦，这是在锻炼你的免疫系统，让你的抵抗能力更强大呢（图4.3）！第

a special weapon (called antibody) to deal with the bad and harmful guys, which is a more efficient way to get rid of the enemy. From this characteristic, scientists invented vaccines. Vaccines are usually made from highly virulent pathogenic microorganisms by decreasing their virulence or after they are killed. When the vaccines enter our bodies, instead of causeing illness, they will stimulate our immune system where the system will go through a smallscale battle practice.

When a second invasion is detected, the immune system will quickly respond to the pathogens and can produce a large number of antibodies targeting the pathogens, and then the pathogens will have

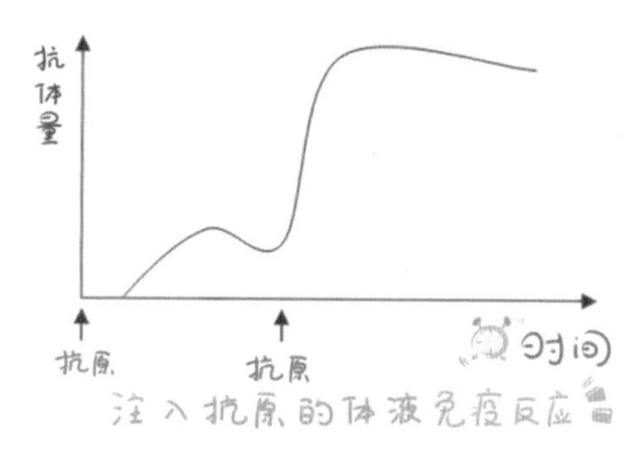

图 4.3　二次免疫反应
Figure 4.3　The secondary immune response

二次注入抗原时，产生抗体的速度大大增快，抗体的量也有很大提升。

接下来，我们就以空气中一个普通的小细菌为例，看看它入侵我们身体的这段旅程中，都经历了些什么。

这天，小细菌和它成千上万的小伙伴们自在地飘在空气中。一阵

to flee. Therefore, when you get vaccinated next time, please do not cry, since the vaccines will make your immune system stronger (Figure4.3)! Both the production rate and the number of antibodies will greatly increase after the second antigen is injected.

Next up, let's take a look at an example of how an ordinary bacterium in the air enters our body, and what it will experience.

One day, after a gust of wind, the bacterium and its thousands of friends in the air floated around a kid, and some bacteria dropped on the kids' skin and tried to enter into the body, but they just stopped at the surface of the firm skin. Some bacteria were

风吹来，它们飞到了一个小朋友身边，有的小伙伴们落在了他的皮肤上，想要钻进去，但皮肤太结实了，它们只能停在表面；有些附在小朋友饼干上被吃进了肚子里，还没来得及得意，就发现这软软的黏膜一点也不好对付，不仅钻不进去，还会分泌一些酸酸的东西，慢慢地，它们被融化掉了；还有的小细菌和其他小伙伴们随着呼吸进入了呼吸道，没想到这里也机关重重，好多小伙伴在半路上就被周围挥舞的小扫把狠狠地甩了出去。小细菌突破重围终于进入了肺里，这时身边的小伙伴已经所剩无几了（图 4.4）！

attached to the kid's cookies and they were eaten into the belly. These little guys figured out that their way was not a smooth ride, as soft mucosa was hard to drill and they were slowly melted by secreted acid. Some bacteria entered into the respiratory tract when the kid is breathing, which was fraught with booby-traps that they had never met before, and many of these guys were dumped out mercilessly. After that, only a few bacteria finally broke through into the lungs (Figure 4.4).

Like a high wall, the skin and mucous membrane are the first defensive line of our body. Their secretions (such as lactic acid, fatty acid, gastric acid and enzymes)

图 4.4 人体第一道防线——细菌被阻挡在皮肤和黏膜外
Fighure 4.4 The first defensive line of body-Bacteria are obstructed by the skin and mucous membrane.

皮肤和黏膜就是我们身体的第一道防线，像高高的城墙一样挡在最外层。它们的分泌物（如乳酸、脂肪酸、胃酸和酶等）就是它们的武器，可以帮助消灭细菌；而呼吸道黏膜上的纤毛，更是能像扫帚一样可以将细菌清扫出去。

穿过肺泡，小细菌进入体液里。这里营养丰富环境舒适，大家立刻

are their weapons that can help us eliminate bacteria. The mucosa cilia of the respiratory tract are just like brooms, they are able to sweep out the bacteria.

After going through the alveoli, the bacteria entered into the body fluid where is nutrient-rich and comfortable, and they began to spread out immediately. But soon they met patrol guards (the lysozymes) who

开始了分裂增殖。可没过多久，它们就遇到了巡逻的哨兵，身边的小伙伴有的被溶解，有的被吃掉，处境越来越危险（图 4.5）。

在这里，细菌遇到的就是我们身体的第二道防线——体液中的杀菌物

dissolve and eat bacteria, and they are then trapped into a dangerous situation (figure 4.5).

Now, the bacteria met the second line of defense—antimicrobial substances and phagocytes in body fluids. The second line can provide non-

图 4.5 人体第二道防线——吞噬细胞的溶菌酶
Figure 4.5 The second defensive line of body—lysozyme of phagocyte

质和吞噬细胞。它们能对外来的可疑目标进行无差别攻击。和皮肤黏膜一样，这道防线也是我们一出生就具备的，对大部分细菌都有防御作用，所以它们都属于“非特异性免疫”。如果这样还是没能拦住入侵的细菌，那就轮到第三道防线出场了。

第三道防线由“免疫器官”和“免疫细胞”两部分组成，是在我们出生后，经过不断训练才逐渐建立起的。它具有只针对某一特定细菌的防御功能，因而叫做特异性免疫。“免疫器官”主要包括扁桃体、淋巴结、胸腺、骨髓和脾，相当于训练士兵的基地；“免疫细胞”小

specific protection against suspicious targets. Just like skin and mucous membranes, this line of defense is given at birth, and it may destroy most of the invaders without targeting specific individual bacteria. Therefore, the first two lines of defense are both non-specific immunity. If that does not stop the invaders, it is then the time for the third line to head into battle.

The third line of defense consists of two parts: immune organs and immune cells, which are acquired through training after birth. They have the ability to target specific germs, so it is called specific immunity. Immune organs, including tonsils, lymph nodes, thymus, marrows and spleens, are like training bases for immune cells.

分队由淋巴细胞、单核/巨噬细胞、粒细胞、肥大细胞等成员组成，它们是攻击的主力。

免疫细胞有两套作战方案：一种主要由B淋巴细胞带领，负责对抗游离在体液里的细菌。B淋巴细胞能识别细菌表面特殊的片段（抗原），并产生一种能够抵抗该细菌的抗体。抗原抗体结合后，吞噬细胞会一拥而上，将抗原清除或使其失去致病性；另一种由T淋巴细胞带领，对付已经在细胞里寄生的细菌。当抗原被检测到时，T淋巴细胞会勇敢地扑上去与被感染的细胞同归于尽。这样一来，混进身体里

Immune cells include leukomonocyte, monocyte/ macrophage, granulocytes, mast cells, and etc., they are the main attackers.

The immunity has two battle plans to fight against bacteria: the first one is called humoral immunity and the other is cellular immunity. Humoral immunity is leaded by B cells, fighting against germs in body fluids. B cells can recognize its matching antigen and synthesize corresponding antibodies. After the binding of antigen to antibody, phagocyte will surround them and wipe them all out. Cellular immunity, however, is led by T cells, and they are mainly fighting against intracellular bacteria. When an antigen is detected, T cells will

的小细菌无处可逃，最终就会被全部消灭掉了（图 4.6）。

和人一样，动物体内也都有这样一套保护自己的免疫系统。哺乳动物的免疫系统和人大致相同。细菌入侵的时候，一样会遇到前面说的三道防线。鸟类有些不一样，它

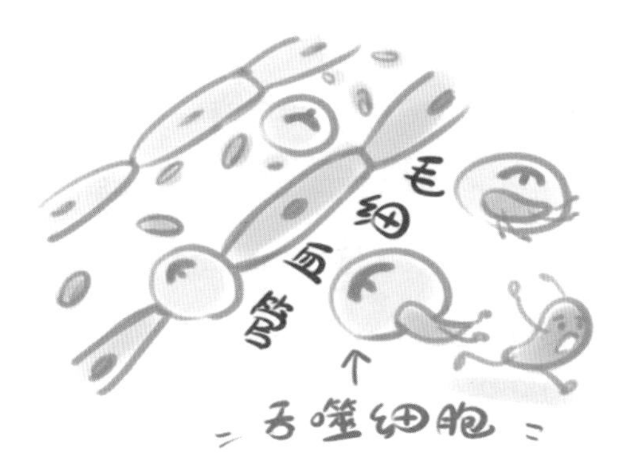

图 4.6　人体第三道防线——特异性免疫细胞针对特定微生物
Figure 4.6　The body's third line of — specific immune cells target at specific germs

gather around to perish together with infected cells. The bacteria, by this time, have nowhere to go and are being defeated eventually (Figure 4.6).

Animals have their self-protecting immune systems just like humans do. Mammals' immune systems are roughly the same as humans'. Bacteria invading into a mammal will come across three lines mentioned above. As for birds, there is something different. Birds have two extra immune organs named the Bursa of Fabricius and the Gland of Harder. To keep animals better protected, veterinarians developed many kinds of vaccines for them too. Therefore, do not forget to take your pets for vaccinations. It is not only for

们多了法氏囊和哈德氏腺两个免疫器官。为了更好地保护它们，动物医生们也研制了许多专门用在动物身上的疫苗。为了家里的小猫、小狗能健康成长，大家也要带它们按时打疫苗哦！

免疫系统从内部作战，默默地保护着我们，消灭掉想要伤害我们的细菌。但有时，细菌的力量太强大了，连最厉害的免疫细胞都无能为力。这时，我们就需要一点外力的帮助啦。抗生素就是人类发现的一种对付细菌的厉害武器。抗生素的发现，让人类和细菌的战争迈入了一个新的历史阶段。

their benefit but for your own health as well.

Our immune systems protect us from pathogens in our bodies, but sometimes due to the high virulence of bacteria, even the most powerful immune cells can hardly do anything against them. At times like this, we need to find some help from outside. Antibiotics are a kind of powerful weapon humans found for ourselves to better fight against bacteria. With the discovery of antibiotics, the war between humans and bacteria escalates to a brand new level.

第五章

抗生素——人类对抗细菌的重要武器

Antibiotics—Powerful Weapons for Humans against Bacteria

1. 抗生素的简史

在不断探索和发现的过程中，我们已经找到了许许多多杀死坏细菌的方法。其中，抗生素对人类和动物来说是最为重要的。抗生素到底是怎么被发现可以用来对抗细菌的呢？这是一段“曲折”的故事。

说到抗生素的发现，一定要感谢一个人，那就是英国细菌学家亚历山大•弗莱明。1928 年，他在培养金黄色葡萄球菌的时候，突然想外出度个假，忘记了实验室在营养培养基上“开心”生长着的细菌。3 周后，当他回到实验室时，注意

1. A Brief History of Antibiotics

Among various kinds of methods for killing harmful bacteria, antibiotics are of prime importance for both human beings and animals. How were they discovered to join the war against bacteria? It's an interesting story.

The discovery of antibiotics is owed to a British bacteriologist, Alexander Fleming. He went on a vacation when he was culturing *Staphylococcus aureus* in 1928 and forgot all about his bacteria, which were growing at a rapid speed on culturing medium. After three weeks, he returned to the lab and noticed a

到在他离开期间，空气中偶然落了一株“奇特的细菌”——青霉菌到培养基上。神奇的事情出现了：在金黄色葡萄球菌的培养基中长出了一团青绿色霉菌，那株“不速之客”的周围呈现出苍白色，在显微镜观察下，青霉菌周围的葡萄球菌都被溶解死亡了（图 5.1）。他认为是青霉菌产生了某种物质，分泌到培养基里抑制了金黄色葡萄球菌的生长。由于当时实验条件艰苦，直到 1941 年，在德国化学家钱恩和英国病理学家弗洛里的共同努力下，真正意义上的青霉素才得以问世，这就是最早发现的抗生素。

strain of Penicillium on the medium. An amazing phenomenon was observed around the colony of the unexpected guest as *Staphglococcus aureus* nearby were dissolved to death under the microscope (figure 5.1). In Fleming's opinion, a certain kind of substance was secreted into the medium and inhibited the growth of *Staphglococcus aureus*. Due to hard experimental conditions, penicillin was not found. It wasn't until 1941 when Penicillin was finally discovered, with the joint efforts from German chemist Chain and British pathologist Florey. This was the first antibiotic discovered in history.

Penicillin finally became the "super medicine" in the World War II fighting against infections. Since it

青霉素成为战场上防治战伤感染的“救命神药”（当时处于第二次世界大战时期）。它大大减少了感染死亡的人数，是第二次世界大战期间十分重要的战略物资。美国把青霉素的研制放在同原子弹的

significantly lowered the death rate in battles, Penicillin was regarded as a vital strategic supply during World War II, and it was even as important as the Manhattan Project in the United States at the time. In 1943, when China was facing the aggression from Japan, a Chinese microbiologist named Zhu

图 5.1 青霉菌杀死附近的金黄色葡萄球菌
Figure 5.1 Penicillium killed the nearby Staphylococcus aureus

研制一样重要的地位上。1943年，中国当时还在抗日后方从事科学研究工作的微生物学家朱既明也分离到了青霉菌，制造出了青霉素，为抗战时期治疗细菌感染做出了巨大贡献。

青霉素这一历史性发现打响了人类与细菌之战的第一枪。科学家们士气大振，在抗生素研发的路上披荆斩棘、奋勇向前。1944年，新泽西大学分离出来第二种抗生素——链霉素。它能有效对抗另一种可怕的传染病——结核。结核在中国以前叫肺痨。我们知道的红楼梦中的“林妹妹”就是因为得了这

Jiming also isolated penicillium in his research and extracted penicillin, which made countless contributions to the war against Japan in China.

The historic discovery of penicillin fired an opening shot in the battle between humans and bacteria. Scientists were greatly inspired and tried to move forward. In 1944, researchers in The College of New Jersey extrated the second antibiotic in human history, streptomycin, which was able to effectively curb a horrible disease-pulmonary tuberculosis. This infectious disease was also called feilao in China decades ago, which led to the death of Lin Daiyu in the classic Chinese novel —— *A Dream of Red Mansions*. The discovery of

种可怕的病而病死的。结核在链霉素发现前几乎没有治愈的可能。这个发现造福了千千万万人。

1947年氯霉素出现。它主要针对痢疾、炭疽杆菌，治疗轻度感染。紧接着，1948年四环素出现，这是最早的广谱抗生素。广谱抗生素的意义在于它能够在还未确诊的情况下“大面积”地杀死很多种类的细菌。时至今日，四环素由于明显的副作用导致四环素牙而基本上只被用于家畜饲养（图5.2）。

1956年，万古霉素诞生，被称为抗生素的最后武器。它可以通过抑制革兰氏阳性细菌的细胞壁合

streptomycin made it possible to cure pulmonary tuberculosis, benefiting thousands of people.

In 1947, streptomycin, aiming at pathogens causing dysentery and anthracnose, were effective for mild infections. Followed by the discovery of tetracycline in1948, which was the earliest broad-spectrum antibiotic in human history, tetracycline is capable of killing many sorts of bacteria before identifying them. Since it has a side effect on our teeth, tetracycline is only used in livestock farming (Figure 5.2).

In 1956, vancomycin was born, and it was supposed to be the ultimate weapon against antibiotics. It has powerful inhibitions on the cell wall synthesis, and it is less likely

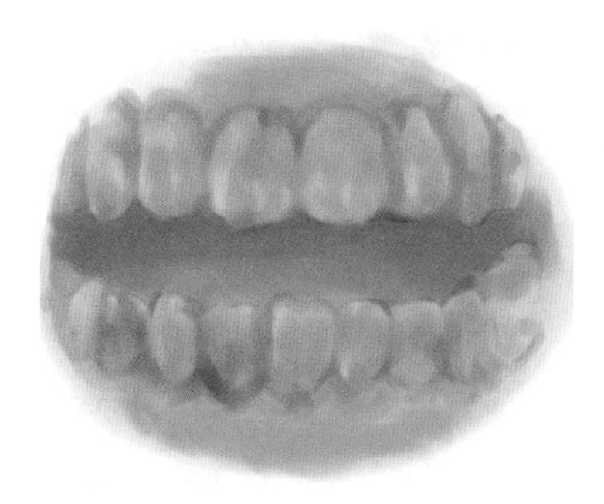

图 5.2 四环素牙
Figure 5.2 Tetracycline stained teeth

成来杀死细菌。杀伤威力强大且不易诱导细菌对其产生耐药性。20 世纪 80 年代，喹诺酮类药物被发现。不同于其他抗生素，它能够破坏细菌染色体，但比较容易引起耐药。

抗生素的发展史其实就是我们

to induce drug resistance. In the 1980s, quinolone was discovered. Being distinct from other antibiotics, quinolone acts on the chromosome of bacteria, and resistance to quinolone is easily developed.

The development of antibiotics is exactly the process of fighting against bacteria. The development of each

不断探索细菌，对抗细菌之路。每一种抗生素是一批又一批科学家智慧和汗水的结晶。这背后都是昂贵的时间和金钱投入。近些年，新抗生素的研制开发速度越来越慢，这就提醒我们必须合理使用已有抗生素。

and every antibiotic not only requires the effort and wisdom of numerous scientists, but also requires a lot of time and money. The discovery of new antibiotics was slow in the recent years, and this is a constant reminder for us to utilize the antibiotics accordingly.

2. 细菌与抗生素之战

2. The Battle between Bacteria and Antibiotics

从青霉素发现到现在，过去了半个多世纪，科学家发现了包括天然（细菌中提取的）和人工合成的近万种抗生素。不过它们之中的绝大多数毒性太大，适合作为治疗人

Since the discovery of penicillin more than half a century ago, thousands of natural and synthesized antibiotics were discovered by lots of scientists. However, most of the antibiotics are highly toxic, only a few

类或动物传染病的药品还不到百种，选择用于人体的抗生素更少。什么样的抗生素才算好的抗生素，能够用于治疗人和动物的细菌感染呢？主要有以下5个条件。

（1）理想的抗生素要对细菌具有选择性的高毒性，而对动物和人体毒性极低甚至无毒。换句话说，我们想要的抗生素不能像“瞎子”一样横冲直撞。不管是好的、坏的、有用的、没用的统统杀死，这样代价太大了。我们希望抗生素可以低毒性。例如，青霉素几乎没有毒性，头孢菌素等毒性也非常低。但青霉素可能会有过敏反应出现。口服或

of them are suitable for the treatment of animal and human diseases. What kinds of antibiotics are good and can be used to treat infectious diseases of human and animals? There are five main aspects.

(1) An ideal antibiotic should be highly effective towards bacteria, but shows low or no toxicity to human and animals. In other words, it can attack the pathogens that cause infections without destroying the other cells. The toxicity of some antibiotics is low, such as penicillin and cephalosporin. Nevertheless, penicillin can cause allergic relations. To ensure safety, the allergic history should be obtained and skin tests should be required before the treatment of patients.

者输液前，要询问患者的过敏史，做皮试等，保证患者的生命安全。

（2）细菌对其不容易产生耐药性。什么是耐药呢？通俗简单地说就是细菌不再“害怕”这种抗生素。为了生存，细菌能够改变自己，最终适应有药物的环境。我们希望耐药性的产生越慢越好，但其实事实不太乐观。1959年科学家发现一种新的半合成青霉素——甲氧西林，在临床上有效地控制了金黄色葡萄球菌的感染，而1961年英国科学家就报道分离到了第一株耐甲氧西林的细菌。世界各地都相继报道了这种耐药细菌引起感染的病

(2) A good antibiotic should not induce antibiotic resistance easily. What is antibiotic resistance? In brief, bacteria develop special ability to prevent themselves from killing by antibiotics. The slower the emerging of antibiotic resistance, the better the antibiotic is. In 1959, a new semisynthetic penicillin – methicillin was discovered and applied to restrain *Staphglococcus* associated infections. Unfortunately, the first methicillin-resistant bacterium was reported by British scientists in 1961, and then infectious cases were reported all over the world. It is terrible that development of a new drug takes decades while drug-resistance can occur in only two years.

例。一种新药需要研制几年甚至几十年，但在两年的临床应用后就产生了耐药性。太可怕了！

（3）适合动物和病人吃药的时间，最好是快速、强效和长效的药物，如一天 3 次或者一天 2 次。

（4）性状稳定，不易被酸、碱、光、热和酶等破坏。抗生素在医院或者药房存放时，如果很容易被外界的东西破坏就很难保存。药物最好可以稳定存放。

（5）使用方便，价格低廉。这应该是患者和生产者的共同愿望吧。

能符合以上要求的抗生素并

(3) The antibiotic should have a rapid onset with potent and long-lasting effects. The time for animals and patients to take medicine can be three times a day or twice a day.

(4) Stability. These antibiotics should not be easily destroyed by acid, alkali, light, heat or enzymes. It is hard to keep antibiotics in a hospital or pharmacy if they are easily destroyed by external factors. It's always the best if medicines can be stored stably for a long period of time.

(5) An ideal antibiotic should be sold at a moderate price and be convenient to take. This is what consumers and producers both wish.

Not many antibiotics can meet all of the requirements mentioned

不多，这些抗生素可以根据“杀伤威力”分为抑制细菌生长繁殖的抑菌剂和杀灭细菌的杀菌剂。不同种类的抗生素对细菌作用的部位也不同，就像不同“武器”具有不同的战斗属性。拥有“重型坦克”性能的抗生素可以破坏细菌铜墙铁壁般的细胞壁；抗生素中的“突击小分队”负责越过并破坏细菌防火线般的细胞膜；而“主力部队”则猛烈的攻击细菌生存的主要物质——蛋白质，最后破坏细菌的指挥中心（核酸）。我们用图来说明这个问题（图5.3）。从图上可以清楚地看到，

above. Antibiotics can be divided into bacteriostatic agents (inhibiting the growth of bacteria) and bactericide (killing bacteria) based on their lethality. Different kinds of antibiotics attack different parts of bacteria, just like different "weapons" possess different attributes. Some antibiotics are like "heavy tanks" and can destroy the cell wall. Some are like "storm troopers" and are responsible for crossing the firewalls and destroying the cytomembrane. Some are the "main force" and can attack protein (essential substance for survival), and finally devastate the nucleic acid(the command center of bacteria). Let's use the following figure to illustrate (Figure 5.3). It is very clear from the figure that

我们常说的青霉素、头孢菌素类都是特异性的针对细菌细胞壁的抗生素。我们也知道我们身体里的细胞没有细胞壁，这也是青霉素和头孢菌素对人体的毒性那么低的原因。

penicillin and cephalosporins, as we have commonly known, act specifically on bacterial cell wall. As cells from our bodies do not have a cell wall, penicillin and cephalosporin are less cytotoxic for human.

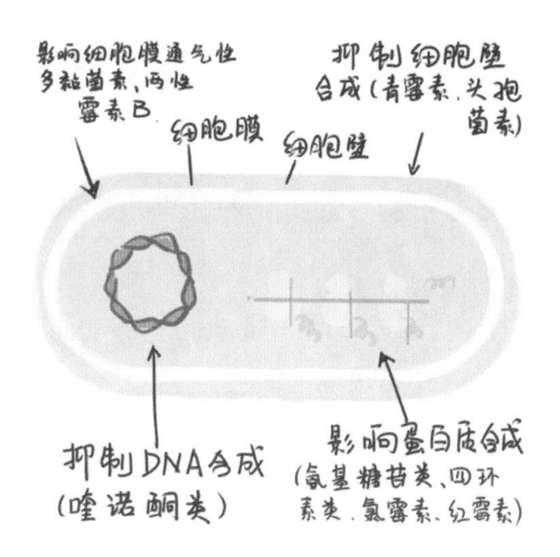

图 5.3 细菌结构与抗生素作用

Figure 5.3 Bacterial structure and antibiotic action

由于引起人类和动物常见细菌感染性疾病的致病菌及其致病方式不完全相同，在选用具体抗生素时存在差异。

As the pathogens causing infectious diseases differ between humans and animals, doctors choose different kinds of antibiotics for disinfection therapy in humans and animals.

2.1 用于动物的抗生素代表

2.1 Veterinary Antibiotics

对付细菌，聪明的人类会采用消毒和灭菌等一系列方法防止坏细菌对我们人体造成伤害。而动物一旦被细菌感染，大多情况下就只能寄希望于我们来帮助它们了。有害的细菌常会使动物表现出呼吸道感染（发烧、咳嗽、流涕）、消化道感染（呕吐、腹泻）、泌尿道感染（尿

Disinfection and sterilization were regarded as advisable methods to kill bacteria to protect us humans. However, if animals are infected with the bacteria they have to rely on humans. Harmful bacteria often attack the respiratory system (causing fever, cough and runny nose), the alimentary tract (manifesting vomiting and diarrhea), the urinary tract (showing

频、尿血）、皮肤感染（红肿、脓疮）等症状，让动物备受折磨。

当动物出现全身发热、咳嗽、打喷嚏、流脓鼻涕等明显的感冒症状，并伴有明显的扁桃体发炎红肿的呼吸道症状时，它们可能患上了细菌性感冒。这种感冒常常是由溶血性链球菌、肺炎链球菌、葡萄球菌或流感嗜血杆菌引起的。我们可以选用人和动物共用的青霉素类抗生素来治疗动物的细菌性感冒。其中，阿莫西林、氨苄西林是比较常用的。

当动物出现高烧、呼吸困难、张口气喘，并从嘴巴和鼻子里流

frequent micturition and hematuria), and skin infection (such as swelling and abscess), which cause animals much suffering.

Animals may suffer from the bacterial colds when showing fever, cough, sneeze and runny tonsil inflammation in respiratory symptoms. The main culprit may be *Streptococcus hemolyticus*, *Staphglococcus pneumoniae*, *Staphylococcus* spp., or *Haemophilus influenzae*. We can choose penicillin shared by human beings and animals to treat bacterial colds. Amoxicillin and ampicillin are commonly prescribed.

Animals may suffer from infectious pleuropneumonia when they show symptoms of fever and respiratory congestion concomitant with red

出红色泡沫样液体等呼吸道感染症状时，它们可能患上了传染性胸膜肺炎。这种疾病常由胸膜肺炎放线杆菌感染引起，有时候是胸膜肺炎放线杆菌与支原体混合感染所致。因此，我们可选用人和动物共用的庆大霉素、卡那霉素、环丙沙星、恩诺沙星等抗生素，也可用动物专用的氟甲砜霉素（商品名纽弗罗）、头孢噻呋钠（商品名沃瑞特）、替米考星等抗生素，或应用金霉素和泰乐菌素（动物专用）来联合治疗。

当动物出现长期腹泻、严重脱水、精神沉郁、食欲降低等消化道

foam flowing from the mouth and nose. This is caused by *Actinobacillus pleuropneumoniae* or mixed infection with mycoplasma. Then we can choose gentamicin, kanamycin, ciprofloxacin and enrofloxacin which are all shared by human beings and animals. Alternatively, aureomycin and tylosin for animals can be chosen.

When animals experience long-term diarrhea, severe dehydration, depression, loss of appetite and other symptoms of gastrointestinal infections, they may suffer from refractory diarrhea. This disease is often caused by pathogenic *Escherichia coli, Salmonella, Proteus* and other *Enterobacteriaceae*, so we often choose the veterinary use

感染症状时，它们可能患上了顽固性腹泻。这种疾病常由致病性大肠杆菌、沙门氏菌、变形杆菌等肠杆菌科细菌引起，因此我们常选用头孢噻呋（动物专用）和头孢喹肟（动物专用）。

当动物出现皮肤瘙痒、严重脱毛，皮肤布满轻重不等的脓疮，并散发腥臭味等症状时，它们可能换上了脓皮症。这种皮肤病常由金黄色葡萄球菌引起，因此我们常选用人和动物共用的头孢拉定、万古霉素、红霉素、青霉素类等抗生素，也可选用动物专用的头孢维星（商品名康卫宁）、羟氨苄青霉素和多

ceftiofur and cefquinome.

When the animals have itchy skin,severe hair loss, and abscess covering their entire body, along with stench and other symptoms, they may suffer from pyoderma. This skin disease is often caused by *Staphglococcus aureus*, so we often use cephradine,vancomycin, erythromycin, penicillins and other antibiotics, which are applicable to both animals and humans.Veterinary antibiotics such as cefavitin (trade name Kang Weining),amoxicillin and polymyxin mixture (trade name Trane) can be alternative choices.

In addition, there are some special bacteria which target particular species of animals. For example,

黏菌素的混剂（商品名牧特灵）等抗生素。

另外，还有一些特殊的细菌，它们主要针对某一类动物。例如，劳森菌，它侵害猪回肠上的黏膜，引发增生性肠炎病，小猪们陆续出现了频繁腹泻、皮肤苍白、消瘦脱水等症状。这时，我们可选用动物专用的沃尼妙林来治疗。例如，引起鸭瘟的默氏杆菌，它侵害鸡、鸭等家禽，使它们出现喘气、咳嗽、下痢甚至神经症状时，我们可选用动物专用的氟苯尼考来治疗。

用于动物的抗生素一般价格都比较低，而且能马上起效。然而，

Lawsonia encroaches the mucosa of the ileum of pigs, causing proliferative enteritis disease. After being infected, piglets exhibit frequent diarrhea, pale skin, weight loss, dehydration and other symptoms.We can use the veterinary valnemulin for treatment. *Riemerella*, which causes duck plague and damages the health of poultry such as chickens, ducks, leading to panting, coughing, diarrhea and even neurological symptoms. To fight against this pathogen, we can use veterinary florfenicol.

The price of veterinary antibiotics is relatively low and they take effect immediately. However, because of their low prices, antibiotics are used as an additive to promote animal's

就是因为价格低廉，很多农场和牧场会预防性地使用抗生素作为添加剂来促进动物的生长。这就造成了农业中抗生素滥用的现象，甚至危害到了我们人类的健康。

growth prophylactically in many farms and ranches, resulting to the abuse of antibiotics in agriculture, which can even endanger our human beings' health.

2.2　用于人的抗生素代表

2.2　Antibiotics for Humans

现在临床用于治疗人体细菌感染的抗生素主要有以下种类。

Nowadays, there are several kinds of antibiotics used for human to fight against bacterial infections.

青霉素类

Penicillins

青霉素类主要是青霉素及和青霉素有同样母核（图 5.4），但改变侧链基团的半合成青霉素。常用的有阿莫西林、氨苄西林、苯唑西林

Penicillins mainly include penicillin and semi-synthetic penicillins which possess the same nucleus (Figure 5.4) with penicillin but differ in the side chain (such commonly used amoxicillin, ampicillin, and oxacillin). "Compound"

等。有时候也制成“复方”制剂，如氨苄西林/氯唑西林，但目前较少用。

我们也知道细菌很聪明。在你想杀死它的过程中，它会想出各种办法把抗生素消灭（这在下面的章节会讲到）。其中一种反抗青霉素的方法就是产生了把青霉素水解掉的酶，我们叫它β-内酰胺酶。于是我们就联合酶抑制剂，来解决这个问题，所以有了青霉素联合酶抑制剂的复方制剂。

preparations, such as ampicillin/cloxacillin, are also included but less used.

We all know that bacteria are very smart.In the process of killing them, they will come up with a variety of ways to eliminate the antibiotics (this will be described in the following chapters). One of the ways to resist penicillin is to produce a group of enzymes which can hydrolyze penicillin (we call them β-lactamases), to cope with the problem, scientists came up with compound antibiotics added with enzyme inhibitors.

头孢菌素类

Cephalosporins

头孢菌素最开始的有效成分来

Cephalosporins came from

源于支顶头孢菌培养液中的一种有效成分。该家族具有同样的母核结构（图5.4）。头孢菌素发展极快，“家族”众多，目前已经有5代头孢菌素了。随着头孢菌素的不断发展，它对人体的肾毒性不断降低，

the active ingredient in the *Cephalosporium acremonium* culture medium. Possessing the same nucleus structure (Figure 5.4), the big family has developed rapidly, so far, there are already 5 generations of *Cephalosporins*. With the continuous development of this antibiotic, the

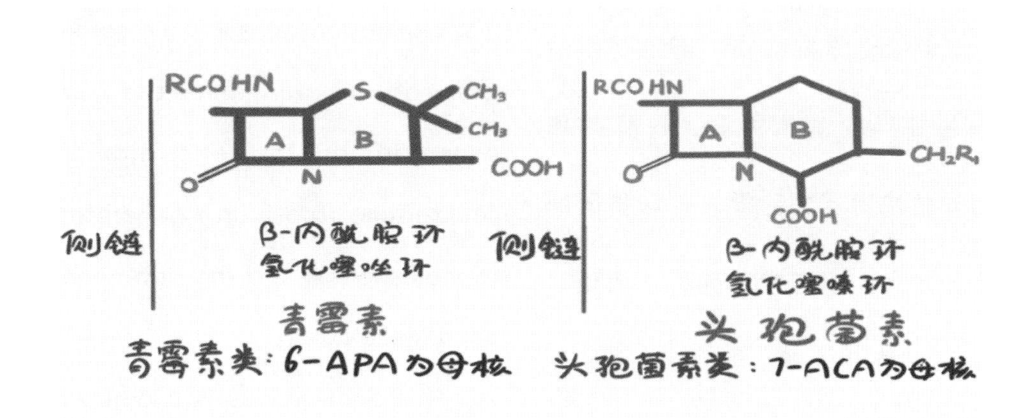

图5.4　青霉素与头孢菌素母核
Figure 5.4　Penicillin and cephalosporin nucleus

对革兰氏阴性菌的作用越来越好，对β-内酰胺酶的稳定性也越来越好（不会被酶分解，就不会产生耐药性）。一些代表性药物有第一代的头孢唑林，第二代的头孢呋辛，第三代的头孢噻肟、头孢曲松、头孢他啶，第四代的头孢吡肟，第五代的头孢洛林。

氟喹诺酮类的作用部位是影响细菌特异的遗传物质。这样的作用部位决定了它的低毒性和抗菌谱广的优点。现在常用的有左氧氟沙星、环丙沙星等。左氧氟沙星还经常作为眼药水的有效成分（图5.5）。

most deleterious side effect of the family, renal toxicity, is getting lower and lower, while the effect on Gram-negative bacteria is getting better and better. Additionally, the stability of cephalosporins to β-lactamases is also getting better and better (without enzyme hydrolysis, bacteria won't develop resistance). Some representatives of drugs are: the first generation cefazolin, the second generation cefuroxime, the third generation cefotaxime, ceftriaxone, ceftazidime, the fourth generation cephalosporins, and the fifth generation cephalexin.

Fluoroquinolones is a group of antibiotics which works on the specific genetic materials in bacteria, and

其他人们常用的抗生素还包括碳青霉烯类（亚胺培南、美罗培南等），氨基糖苷类（庆大霉素等），四环素类，糖肽类（万古霉素、替考拉宁等）等。还有专门用于厌氧菌感染治疗的硝唑类和抗真菌药物（康唑类等）。

抗生素不同的作用部位也提醒我们要针对细菌有选择性地使用合

图 5.5　左氧氟沙星滴眼液
Figure 5.5　Levofloxacin eye drops

is of low toxicity with broad antibacterial spectrum. The ones commonly used today include levofloxacin, ciprofloxacin, etc. Specifically, levofloxacin is often used as the active ingredient of eye drops (Figure 5.5).

Additionally, other commonly used antibiotics include carbapenems (such as imipenem, meropenem), aminoglycosides (such as gentamicin), tetracyclines, glycopeptides (such as vancomycin, teicoplanin), azole (specific for anti-anaerobes) and antifungal drugs (such as fluconazole).

Distinct mechanisms of antibiotics remind us to selectively use appropriate antibiotics against specific bacteria, rather than blindly

适的抗生素，而不是盲目用药。细菌与抗生素的对抗之路漫长曲折。“道高一尺，魔高一丈”，抗生素耐药细菌的出现速度甚至超过了新药研发的速度。随着抗生素的滥用现象日渐严重，耐药性不断上升，合理使用抗生素的呼声越来越高。

apply medication. There is a long road ahead for the fight between bacteria and antibiotics. As virtue rises one foot, vice rises ten, the emergence of drug-resistant bacteria even exceeds the speed of drug development. With the severity of antibiotics abuse,drug-resistance is on the rise, the plea for reasonable use of antibiotics is getting more and more urgent.

第六章

适者生存的产物
——耐药细菌

Survival of the fittest
—Drug-resistant bacteria

1. 抗生素耐药细菌的出现

抗生素成为了细菌的头号大敌。然而“道高一尺，魔高一丈”。1945年，英国科学家亚历山大•弗莱明因为发现了青霉素而获诺贝尔生理学和医学奖。当人们还沉浸在自以为拥有青霉素就可以完败细菌而高枕无忧时，这位智者就高瞻远瞩地提出了会出现耐药细菌的预测。果不其然，被誉为“抗生素的黄金时代”的20世纪，见证了一种又一种新的抗生素的诞生，同时也见证了耐药菌的出现。

1. The Emergence of Antibiotic Resistant Bacteria

Antibiotics have become the principal enemy of bacteria. However, as the saying goes: "As virtue rises one foot, vice rises ten". In 1945, British scientist Alexander Fleming won the Nobel Prize in Physiology or Medicine for the contribution of discovering penicillin. When people were still immersed in the belief of penicillin as the ultimate protection against bacteria, Fleming made a far-sighted prediction of the emergence of drug-resistant bacteria. As expected, the 20th century that was known as the "golden age of antibiotics", had witnessed the birth of many new

青霉素是人类发现的第一种抗生素。青霉素耐药的细菌直到20年后才姗姗来迟。1959年，对仅仅出现一年的四环素就报道出现了耐药的志贺菌。几年后，耐甲氧西林金黄色葡萄球菌（MRSA）也横空出世。这种耐药菌到现在还是让医生们头疼不已。1953年，红霉素开始用于救治病人，15年后红霉素耐药的链球菌出现。庆大霉素耐药的肠球菌也是在庆大霉素问世后十几年的时间内出现的。

1972年，被誉为“MRSA克星”的万古霉素出现。这让人们欢欣鼓舞，然而万古霉素耐药的肠球

antibiotic, as well as the emergence of drug-resistant bacteria.

Penicillin was the first kind of antibiotics discovered by human, however, penicillin-resistant bacteria only arrived 20 years later. In 1959, it was reported that the tetracycline-resistant *Shigella* was found only one year after the discovery of the tetracycline. A few years later, Methicillin-resistant *Staphylococcus aureus* (MRSA) emerged, causing problems for doctors even today. In 1953, doctors began to use erythromycin to treat patients, but erythromycin-resistant *Streptococcus* appeared after 15 years. Gentamycin-resistant *Enterococcus* also appeared within two decades after the application

菌（VRE）和万古霉素耐药的葡萄球菌（VRS）也接踵而至。亚胺培南是碳青霉烯类抗生素家族中的成员，被称为对抗耐药的革兰氏阴性菌的“金牌卫士”。不幸的是，碳青霉烯类抗生素耐药的肠杆菌科细菌（CRE）早已出现，CRE队伍不断壮大，甚至在全球范围内扩张。

看到这里，你们是不是会觉得这些耐药细菌似乎总是能够克服抗生素，找到自己的生存之路？这些耐药细菌到底是怎么出现的呢？科学家们发现，耐药细菌的出现和抗生素的长期广泛使用，尤其是抗生素滥用密切相关。当抗生素大规模对

of gentamycin.

In 1972, the isolation of vancomycin which was dubbed "the bane of MRSA" was a reason for celebration, but alas vancomycin-resistant *Enterococcus* (VRE) and vancomycin-resistant *Staphylococcus aureus* (VRSA) appeared not long after. As a member of carbapenems, imipenem is regarded as the "golden guard" to fight against gram-negative bacteria. Unfortunately, carbapenem-resistant *Enterobacteriaceae* (CRE) has already been reported.

It seems that the resistant bacteria can always overcome antibiotics and find ways to survive. So, how did these resistant bacteria come about? Scientists found that the emergence of drug-

细菌产生作用时，大部分细菌不能承受强大的杀伤威力而死亡，只有极少一部分细菌还在垂死挣扎。如果抗生素浓度不足以杀死全部细菌，这些狡猾的细菌就会慢慢地重新武装自己以适应恶劣的环境，并存活下来。相比于一般的细菌，这些耐药菌就像打不死的“小强”，常常具有更顽强的生命力，更难被消灭。

真正令人担心的是，这些“顽固分子”身上携带的“秘密武器”——耐药基因能潜入其他种类的病菌并相互传递蔓延。这样一来，很多敏感菌株也变得耐药，耐药性就传播开来了（图6.1）。

resistant bacteria is correlated with the extensive use of antibiotics, especially the abuse of antibiotics. Most bacteria die under the effect of a large dose of antibiotics, with very few survivors. If the concentration of antibiotic is inadequate to kill all of them, the cunning survivors will adapt to the hostile environment. Compared with common bacteria, drug-resistant bacteria have much stronger vitality and are harder to be killed.

What's worse, the drug-resistant genes, carried by these bacteria as secret weapons,can be transferred to other kinds of pathogenic bacteria and spread. In this way, a large amount of previously sensitive strains gradually become resistant to antibiotics (Figure 6.1).

图 6.1　耐药细菌与抗生素之战
Figure 6.1　The battle of drug-resistant bacteria and antibiotics

2. 耐药机制——细菌对抗抗生素的秘密武器

我们都知道，青霉素是青霉菌产生的。其实，大多数的天然抗生素都是微生物抵御外敌入侵保护自身安全的代谢产物。科学家们“以

2. Resistance Mechanisms—The Secret Weapons of Bacteria Confronting Antibiotics

As we know, penicillin was produced by *Penicillium*. In fact, most natural antibiotics are metabolites

其人之道，还治其人之身”，成功地把这些抗生素提取出来用来对付危害人类健康的坏蛋细菌。但是一些狡猾的细菌拥有秘密武器来抵抗抗菌药物。我们来看看他们都有哪些秘密武器吧。

首先，是灭活酶的产生。这是细菌最重要的秘密武器之一。灭活酶，顾名思义，就是能够灭活抗生素，使它们对细菌不能发挥杀伤威力的酶。这类酶是一个大家族，包含好多成员。其中，β-内酰胺酶家族“人丁兴旺”，迄今为止超过300种。β-内酰胺酶能使β-内酰胺类抗生素的β-内酰胺环打开，

from microbes to protect themselves from their enemies. Scientists extract antibiotics to fight against harmful bacteria. However, some “special strains”develop secret weapons to protect themselves from antibiotics. Let’s take a look at these weapons.

First, inactive enzymes, one of the most important secret weapons of bacteria, can inactivate antibiotics, rendering antibiotics useless. These enzymes form a large family, consisting of many members, such as β-lactamases, which contain over 300 species. β-lactamases can inactivate β-lactam antibiotics by cutting β-lactam rings (Figure 6.2).

从而使抗生素失效（图 6.2）。

现在让医生们和病人们都很头疼的有超广谱 β- 内酰胺酶（ESBLs）、头孢菌素酶、金属酶等。这些讨厌的家伙的存在使青霉素、头孢菌素这些 β- 内酰胺类抗生素的治疗效果越来越差。其中，金属酶是威力最强的 β- 内酰胺酶，能够破坏几乎所有的 β- 内酰胺类抗生素。2009 年，产 NDM-1 金属酶的超级细菌的发现引起全球轰动，人们谈“菌”色变。

此外，还有一些酶，如氨基糖苷修饰酶，扮演着“化妆师”的角色。它们能够给抗生素“乔装打扮”，

Some of them, including extended-spectrum β-lactamase (ESBLs), cephalosporin enzymes and metal enzymes, are troublesome as they can invalid penicillin and cephalosporin. Among them, metal enzymes are the

图 6.2 细菌产生的灭活酶破坏抗生素结构使抗生素失效

Figure 6.2 Antibiotics are inactivated by bacteria who produces inactive enzymes to destroy antibiotics structure.

由于被改头换面的抗生素不能与细菌的活性部位接头而失去活性，所以氨基糖苷修饰酶也是灭活酶的一种，在细菌的耐药过程中起推波助澜的作用。

值得一提的是，编码这些灭活酶的基因可“落户”在染色体或质粒上。位于染色体上的耐药基因能够从“爸爸”到“儿子”再到“孙子”代代相传，而位于质粒上的基因能够借助质粒这条自由穿梭的小船在细菌家族成员之间互相传播。因此，大范围播散的耐药菌，它们的耐药基因大多位于质粒上。

改变药物作用的靶位也是这些

most powerful weapon, which can destroy all of β-lactam antibiotics. NDM-1-harboring superbugs were discovered in 2009, which became a sensational news, and people were trapped in panic when talking about "bacteria".

Besides, some enzymes, like aminoglycoside-modifying enzymes, play a role in modifying antibiotics and leading to the invalid combination between antibiotics and the combining site on bacteria. Since antibiotics loss their activity in the situation, the enzymes are named inactive enzymes.

It is worth mentioning that these enzyme-encoding genes can be located on chromosome and/or plasmid. Chromosomal genes can be transmitted

狡猾的耐药细菌的惯用伎俩。抗生素首先要与细菌身上的某些特殊的蛋白结合才能进一步发挥杀伤作用。一旦改变这些结合部位的蛋白，抗生素就不能与细菌好好结合，抗菌作用肯定会受到影响。

例如，肺炎链球菌对青霉素的耐药性就是因为青霉素结合蛋白改变，导致青霉素不能很好地结合到细菌上，从而使细菌躲过青霉素的追杀。有的细菌与抗生素接触后还能产生新的靶蛋白，混淆视听从而躲过一劫。例如，科学家们在MRSA身上发现的不同于敏感的金黄色葡萄球菌的青霉素结合蛋白的

from generation to generation, while the genes on plasmid can be transferred among bacteria. So, resistance genes of pandemic resistant bacteria are mostly located on plasmids.

Changing target sites of antibiotics is another trick of these cunning resistant bacteria. Antibiotics need to combine with some special protein on the bacteria to kill them. Once these protein-binding sites change, antibiotics cannot combine with the bacteria, thereby losing its potency.

For instance, the change of penicillin binding proteins (PBP) leads to the penicillin-resistance of *Streptococcus pneumoniae*, resulting in the invalid combination between antibiotics and bacteria. To survive,

青霉素结合蛋白2a（PBP2a）。

除了灭活酶和药物靶位的改变外，细菌外膜通透性和外排泵系统的改变也是负隅顽抗的细菌耐药性形成的策略之一。因为细菌表面有一些由OmpF和OmpC膜蛋白组成的通道。这些通道能允许抗生素等小分子从高浓度的体外自由进入体内。当它们与抗生素接触一段时间后，有的“投机分子”就会发生突变，使OmpF通道蛋白丢失，于是进入体内的抗生素就大大减少。另一方面，有些细菌能够消耗能量，有针对性地把进入体内的多种抗生素通过类似“水泵”的外排泵系统

some bacteria can produce new target proteins to fool antibiotics. For example, PBP2a, being different from PBP of sensitive *Staphylococcus aureus*, was discovered in methicillin-resistant *Staphylococcus aureus* (MRSA).

Apart from inactive enzymes and changing of drug targets, the modifying of outer membrane permeability and efflux pump system can also cause antibiotic resistance. There are some channels composed of OmpF and OmpC membrane protein existing on the surface of bacteria which allow small molecules such as antibiotics get in. Exposed to antibiotics for a while, some bacteria may discard OmpF, thus lessening the

排出体外。常见的有大肠埃希菌中的AcrAB-TolC系统和铜绿假单胞菌的MexAB-OprM系统。它们使大肠埃希菌和铜绿假单胞菌对多种药物耐药。

“兄弟同心，其利断金”的道理在微生物王国同样是真理。对于单个的细菌，抗菌药可能很快就能让它走投无路缴械投降，但一旦它们齐心协力筑成坚固的“城墙”——生物被膜，这场战争就没那么容易分出胜负了。科学家们发现，有了这道城墙，细菌的耐药性可大大增强。抗生素非但不能有效地清除生物被膜，还会诱导耐药性的产生。

amount of antibiotics that can get into these bacteria. On the other hand, some bacteria can pump antibiotics out of their body. The most common efflux pump system are AcrAB-TolC of *Escherichia coli* and MexAB-OprM of *Pseudomonas aeruginosa*, which lends to multi-drug resistance of these bacteria.

As the saying goes, "brothers of the same mind, their sharpness can cut through metal", it is also applicable in the microbial world. Antibiotics can destroy a single bacterium, but not the bacterial biofilm formed by a large number of bacteria. With this biofilm, resistance of bacteria is significantly enhanced. Antibiotics cannot effectively remove

正是因为这些秘密武器，耐药细菌在克星抗生素的全力围剿下还能负隅抵抗，甚至扳回一局（图6.3）。

this biofilm, and may actually induce the onset of resistance.

In the war with antibiotics, drug-resistant bacteria can survive owing to their secret weapons, and can even win some battles (Figure 6.3).

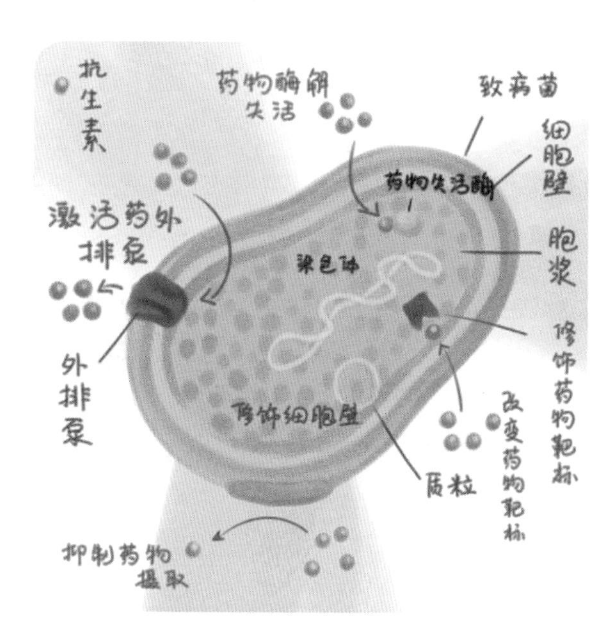

图 6.3 细菌耐药的秘密武器——改变自身结构

Figure 6.3 The secret weapons of bacteria—changing its own structure to defense antibiotics and produce resistance

3. 耐药细菌军团日益壮大

正如歌词“what doesn’t kill you makes you stronger”所唱的，没能打倒你的对手将会使你变得更强。细菌与抗生素之战中存活下来的耐药菌，非但没有被消灭，反而日益壮大。

国内最权威的细菌耐药监测网CHINET 2005—2014年连续十年的数据告诉我们，在病人身上分离的肠杆菌属细菌中，有超过90%的菌株对头孢唑林和头孢西丁耐药，超过10%的菌株对酶抑制剂复方制剂显示耐药。大肠埃希菌是最常

3. An Increasing Number of Drug Resistant Bacteria

As the lyrics go, "what doesn't kill you makes you stronger". Bacteria that survive from the war against antibiotics have become more and more powerful.

According to the statistics of the last decade (2005—2014) from CHINET, the most authoritative antimicrobial resistance surveillance system in China, over 90% of the *Enterobacteriaceae* isolated from clinical patients are resistant to cefazolin and cefoxitin, while more than 10% of them show resistance to enzyme inhibitor compound. As for *Escherichia coli*, one of the most common pathogens, the ten years' data

引起感染的病原菌之一。科学家们十年监测发现，大肠埃希菌中超广谱 β-内酰胺酶 ESBLs 的检出率不断上升。这些搞破坏的酶的存在使大肠埃希菌能够抵抗许多抗生素。

在引起人们血流感染的革兰氏阴性病菌中，肺炎克雷伯菌排名第二。十年的研究发现，克雷伯菌属对碳青霉烯类药物耐药率不断升高。2005 年，只有 2.9% 的克雷伯菌属细菌对亚胺培南耐药，而十年后这个耐药比例超过了 10%。这个趋势对我们来说非常不乐观。类似地，克雷伯菌属菌株对美罗培南的耐药率，也从 2005 年的 2.9% 升

indicates that the detection rates of ESBLs (extended spectrum β-Lactamase) among *Escherichia coli* have been increasing continuously, leading to wide resistance to many antibiotics.

Klebsiella pneumonia ranks second in the gram-negative bacteria causing blood infections, and the decade study also found increasing carbapenem resistance. To be specific, only 2.9% of *Klebsiella* was resistant to imipenem in 2005, however, in ten years, the resistance rate rose to over 10%, which is very grim indeed. Similarly, the resistance rate of meropenem had risen from 2.9% to 13.4% during the period from 2005 to 2014. What extremely worries us is the fact that more than 60% of carbapenem resistant *Klebsiella*

至2014年的13.4%。尤其让我们担心的是，碳青霉烯类抗生素耐药的克雷伯菌属对大多数其他抗菌药的耐药率在60%以上。这类细菌到最后基本无药可用。

抗生素耐药的超级细菌对人类的严重威胁已经蔓延至全世界所有国家和地区，影响着地球上的每一个人。2014年，世界卫生组织WHO发布史上第一份全球抗生素耐药监测报告。报告指出对碳青霉烯类抗生素耐药的超级细菌已经传播到全世界各个地区。碳青霉烯类抗生素作为对抗严重感染的最后一道防线已被突破。有的国家由于

shows resistance to other antimicrobials, leading to no effective treatment.

Superbugs that are resistant to antibiotics are threatening everyone everywhere. In 2014, WHO issued the first-ever global report, a surveillance on antimicrobial resistance, the report revealed that superbacteria resistant to carbapenem has spread all over the world, and carbapenem being the last resort in fighting against serious infections had been losing its effectiveness. In some countries, due to high resistance, carbapenem are invalid for over half of the patients infected with *Klebsiella pneumoniae*. Fluoroquinolones, the most commonly used treatment of *Escherichia coli* associated urinary tract infections,

耐药率高，碳青霉烯类抗生素对半数以上接受治疗的肺炎克雷伯菌感染的病人无效；氟喹诺酮类药物是最广泛用于治疗大肠杆菌引起的尿道感染的抗菌药物之一。同样，某些国家和地区，这种药物对半数以上的病人无效；世界上每天有超过100万人感染淋病。第三代头孢菌素被用作淋病最后的治疗手段，然而奥地利、澳大利亚、加拿大、法国、日本、挪威、南非、斯洛文尼亚、瑞典和英国等国家都有治疗失败的病例。

前英国首相卡梅伦委托著名经济学家Jim O' Neill调查超级菌对

are invalid for more than half of the patients in some countries and areas. More than one million people are suffering from gonorrhea around the world every day, and the third generation cephalosporins have been considered as the last resort of treatment for gonorrhea. However, failure cases have been confirmed in several countries, including Austria, Australia, Canada, France, Japan, Norway, South Africa, Slovenia, Sweden and the United Kingdom.

The renowned economist Jim O'Neill was commissioned by the former British Prime Minister David Cameron to investigate the influence of superbacteria in various fields over the world, and the results were

世界各国各方面的影响，结果令人大吃一惊。根据统计，现在全球每年因为耐药微生物感染死亡的人数约为70万人。如果再不采取有效措施的话，这一数字到2050年将超过1000万人，甚至超过癌症的死亡人数。仅在欧洲和美国，每年就有约5万人因为感染超级细菌而死亡。到2050年，这一数字将变成现在的十倍。由此而产生的花费也将达到惊人的63万亿英镑（英国目前的GDP也仅为3万亿英镑），占到了全球经济总量的2%~3.5%。

不管是耐药菌感染所致的死亡

astonishing. According to statistics, at least 700000 people died from antimicrobial-resistant infections each year. Furthermore, more than 10 million people would die in 2050 if we do not take effective action, it will exceed the death toll caused by cancer. In Europe and the United States alone, around fifty thousand people die each year due to superbacteria infections. In 2050, this number will be ten times the number today, and the cost will amount to sixty-three trillion pounds (British current GDP is three trillion pounds), taking up 2%~3.5% of total global GDP.

Whether it's the death toll or the economic losses caused by antimicrobial resistance infections, these statistics

人数，还是抗生素耐药造成的经济损失，这些触目惊心的数据，给人类敲响了警钟。耐药军团日益庞大，而抗生素步入穷途末路，后抗生素时代即将到来。人类将何去何从？

have set off an alarm bell for humanity. The increasing number of antibiotic resistant bacteria, and the low-effectiveness of antibiotics indicate that the post-antibiotics era will arrive. How can humans cope?

第七章

应对细菌耐药的挑战，我们可以做什么

What Can We Do for Tackling Antimicrobial Resistance

耐药细菌的队伍越来越壮大，而能够有效对抗这些超级细菌的抗生素越来少。后抗生素时代的阴影挥之不去，人类健康的未来也岌岌可危，那我们是不是就只能坐以待毙？当然不是！在这场没有硝烟的战争中，我们又能做点什么来积极应战呢（图7.1）？

首先，我们必须要充分地认识到抗菌药物滥用的巨大危害。现在在很多情况下，由于我们合理用药的意识不强烈，尤其是抗生素的不合理使用，引起了更多耐药的超级细菌出现。这些细菌像是穿了盔甲的战士，使原本有效的抗生素的效

With a growing number of drug-resistant bacteria and a decreasing amount of effective antibiotics, the post-antibiotics era seems to be right around the corner. The public health of our society will be on the ropes if we do nothing except waiting for our doom. So, what can we do facing this invisible war against super bacteria (Figure 7.1)?

First, it is vital for us to be aware of the great danger in abusing antibiotics. Under many circumstances, the emergence of superbacteria results from a lack of appropriate drug-use, especially the inappropriate use of antibiotics. These new bacteria possess drug-resistance mechanisms which reduce or even eliminate the

图 7.1　后抗生素时代，超级细菌
Figure 7.1　Post-antibiotics era and super bacteria

果降低了，甚至完全不起作用。这些坏蛋就像定时炸弹，随时爆炸，我们的健康岌岌可危。

effectiveness of antibiotics, although antibiotics were deadly to bacteria previously.Thus, this situation puts public health under great danger.

因此，不仅自己要有合理使用抗生素的想法，同时作为活跃的传

Consequently, we should not only enhance our awareness of using

声筒，我们还要积极地向身边的家人、老师、朋友宣传“杜绝抗生素滥用，坚持合理使用抗生素”的理念。只有大家都意识到抗生素滥用的危害性，才能真正从源头上减少抗生素滥用。

其次，先下手为强——预防感染，不失为明智之举。在细菌没有引起我们生病时，我们一定要多运动、多锻炼、强身健体，增强我们的抵抗力，由内到外武装自己，让这些耐药菌军团无从下手。对于抵抗力比较差或者已经生病的人来说，注射疫苗是一个主动提高自身免疫力的好方法。

antibiotics reasonably, but also impart this consciousness to people around us, such as our friends, teachers and family members. Only if all of us realize the hazard of antibiotics abuse, can we cut down this pervasiveness problem from the source.

Another effective method is preventing infection. While we are healthy, we need to take the initiative to strengthen our immune system like participating in outdoor activities more frequently and adopting a balanced diet to prevent bacteria from infecting us. For those with weak immune systems or are already sick, vaccination is an acceptable method to boost their immunity.

近年来，“耐药细菌的克星”——超级细菌疫苗的研发被WHO（世界卫生组织）、欧美国家政府及医药巨头公司所重点关注。国内外的科学家们已经研制出多个金黄色葡萄球菌疫苗、铜绿假单胞菌疫苗和鲍曼不动杆菌疫苗。这些疫苗已经进入不同的临床试验阶段，相信不久的将来肯定可以应用到病人身上。它们能够诱导我们免疫系统有针对性地对细菌的感染进行反抗，就像射箭一样，能针对性地射击靶心，减少这些细菌的感染，减少抗菌药物的使用，缓减耐药菌的出现，被科学家们寄予了厚望

In recent years, the Preventive Killer-vaccines aiming at super bacteria has been put under the spotlight by the WHO, western governments and medical corporations. Scientists have developed various kinds of vaccines against *Staphylococcus aureus*, *Pseudomonas aeruginosa* and *Acinetobacter baumannii*, most of which are in different stages of clinical trials. They will likely be put into practical use in the foreseeable future which can induce our immune system to fight against infections in the way that targets right at the "bull's eye" of the bacteria and causes less selective pressure and thus slows the appearance of drug-resistant bacteria

图 7.2 疫苗帮助人们对抗超级细菌的入侵
Figure 7.2 Vaccine against super bacteria

（图 7.2）。

对于科学家们来说，只有与时间赛跑，加紧研制开发“武器”——新抗生素，才能在这场无形的战争中把握主动权。近些年来，全球多种抗菌药物得到相关部门的认证进入我们的视野，包括难辨梭

(Figure 7.2).

To get the initiative in this invisible war, new antibiotics are urgently needed. In recent years, a number of antibacterial agents have been certified by relevant departments including Cadazolid, a protein synthesis inhibitor of *Clostridium difficile*; Dalvance, the

状芽孢杆菌中蛋白质合成抑制剂Cadazolid，针对性的控制难辨梭菌的感染；第一个适用于急性细菌性皮肤、皮肤结构感染治疗双剂量方案静脉注射抗生素——半合成脂糖肽Dalvance；用于淋病治疗的抗生物AZD0914；Insmed公司旗下的吸入式抗生素药物Arikayce；另一用于急性皮肤及其附属结构感染的新型治疗抗生素Sivextro（该药可对耐甲氧西林金黄色葡萄球菌和链球菌属等多种细菌起到抗菌作用）；Medicines旗下敏感革兰氏阳性菌抗生素Orbactiv；以及腹腔感染、尿路感染治疗用抗生素Ceftazidime-

first intravenous antibiotic which can be applied to both bacterial skin infections and structural skin infections in a double-dose method; AZD0914, an antibiotic which is effective to treat gonorrhea; Arikayce, an antimicrobial inhalation developed by Insmed; Sivextro, a new antibiotics for acute infection of skin and its accessory structure, active against MRSA and *Streptococcus*; Orbactiv, which focuses on gram-positive bacteria developed by Medicines company; Ceftazidime-avibactam, aims at peritoneal infection and urinary tract infection. These new "weapons" are attributed to the great efforts from scientists who are constantly devoting themselves to the development of novel antimicrobial

avibactam。这些“新武器”凝聚了很多人的心血。除此之外，还有许许多多的抗菌药物在科学家们的辛勤努力下处于实验室研发阶段。

每一种新的抗生素的诞生，都需要科学家们数年、数十年的精力投入，耗资巨大。更要命的是，耐药菌的出现猝不及防。在这样的情况下，合理地利用已有的抗生素显得尤为重要。我们都知道，没有细菌感染症状就预防性地用药，没有根据病原体选择针对性的抗菌药物，药量不足，该停药的时候不停药，不该停药的时候停药，这些错误的行为都会导致耐药细菌的出现

medicines. In addition, under the hard efforts of many scientists, there are many antibiotics being researched in labs.

Each discovery of a new antibiotic costs scientists countless resources both in time and money. It usually takes decades to develop a new drug. However, drug-resistant bacteria appear at a much faster speed. Thus, the appropriate use of existing antibiotics is of great importance. Unreasonable therapies include taking antibiotics for preventive purpose, using antimicrobial agents without a targeted spectrum of the pathogen, insufficient dosage and shortening or prolonging medication time all lead to the occurrence of drug-resistant

图 7.3　抗生素滥用
Figure 7.3　The abuse of antibiotics

(图 7.3)。

医生们需要借助实验室里快速、准确的诊断方法来抓住病原菌。只有通过病原菌对药物的敏感性试验，才能及时选择有效的

bacteria(figure 7.3).

To avoid the appearance of drug-resistant bacteria, it is sensible for doctors to identify pathogens through accurate laboratorial methods and choose appropriate antibiotics according to the results of antimicrobial

抗生素。这样就能大大减少耐药菌的出现。这也是实验室一项最重要的职责。而一旦耐药菌出现，医生们就要当机立断，把携带耐药菌的病人转移到隔离病房，同时还要做好病房环境的消毒工作，防止这些“恐怖分子”通过一系列传播途径，尤其是病房里一些物体表面，手的接触等，传播到别的病人身上。所以勤洗手对于防止病菌传播也是很有用的。

总而言之，为了能够有效控制和延缓细菌的耐药，我们要贡献上我们每个人的力量。人们必须转变错误的用药观念，纠正错误

susceptibility tests. Patients should be isolated immediately into disinfected wards and strengthening work of ward disinfection once confirmed infection. It's important to cut off transmission route, especially bacteria sticking on surfaces or easily spreading by hands contact. Washing hands is another effective way to prevent transmission of bacteria through poor hygiene.

In conclusion, to curb the spread of drug-resistant bacteria, we must all do our part. People need to take a reasonable attitude towards antibiotic therapies, correct the inappropriate medicine utilization behavior, abandon the wrong habit of medicine use, and reduce the unreasonable use of antibiotic medicine.

的用药行为，摒弃错误的用药习惯，切切实实减少抗菌药物的不合理使用。

（1）医务工作者应肩负起控制感染的重大使命。应坚持因病施治，以药敏作为依据减少经验用药，严格掌握药物的适应症，并首先把控并指导患者合理用药。

（2）通过多种渠道进行医疗宣教，使大家认识到抗菌药物滥用的巨大危害，从而不自行购买和盲目使用抗菌药物。

（3）纠正错误的用药观念。很多人一直坚信昂贵的新药必定是疗

(1) Medical workers should shoulder the important mission of infection control. They should apply treatments which aim accurately at the pathogen based on antimicrobial susceptibility tests instead of empiricism.

(2) Increasing awareness of the undesirable consequences from antibiotics abuse through multiple channels so that the public will stop purchasing and blindly using antibiotics.

(3) Correcting mistaken medication concepts. Many people believe that costly new medicines must be the most effective, therefore, they push doctors to prescribe the most advanced broad spectrum antibiotics

效最为显著的，极力要求医师给自己用高级广谱的抗菌药物，甚至认为抗菌药物为万能的“消炎药”，连无菌性炎症也选择使用，导致抗菌药物的滥用，甚至还造成药源性疾病的发生（图 7.3）。

（4）严格控制处方药销售。抗菌药属处方药，应当凭医师处方销售。药品零售企业和一些小药店、小诊所不要为了追求经济利益擅自无处方售药或盲目推荐新药和贵重药，要严控自己的行为。

（5）规范化管理。严格执行国家出台的有关抗菌药物专项治理规

and even regard them as all-purpose "anti-inflammatory medicine", going so far as to use them even for aseptic inflammations, leading to the abuse of antibiotics and even occurrences of drug-induced diseases (Figure 7.3).

(4) Strict control of prescription sales. Antimicrobials are prescription drugs that should be sold with physician prescription.Pharmaceutical retail corporations, pharmacies, and clinics should not sell drugs without prescription or recommend costly newer drugs blindly for pursuing economic gain. Instead, they need to strictly control their behaviors.

(5) Standardized management. Standardizing the application of antibiotics. Strictly following the

定及相关法律、法规和规章，加强无菌操作技术规范，规范抗菌药物临床应用行为，控制细菌耐药，保障医疗质量和确保广大人民群众的用药安全。

regulations regarding antibiotics from authorities and guaranteeing the security of the public; strengthening the technical specifications of aseptic operations;regulating the clinical application of antimicrobial agents;controlling bacterial resistance in order to protect the quality of medical care and ensure the safe application of drugs for the general population.